International Journal of Research (IJR) ISSN: 2348-6848

INTERNATIONAL JOURNAL OF RESEARCH (IJR)

VOL-1, ISSUE-3

Publication

This journal is published monthly from

SureShotPOST Online Publishing

CG-14, Jawahar Bhawan, Indian Institute of Technology, Roorkee, Pin-247667, India

And

B-305, Prem Nagar, Suleman Nagar, Korari, New Delhi, Pin-110086, India

Editorial Board

For advertising and sending articles for publication you can contact

Editor-in-Chief

International Journal of Research (IJR)

CG-14, Jawahar Bhawan,

Indian Institute of Technology, Roorkee

Pin-247667

Uttrakhand, India

Contact No. +919958037887

Email: contact@internationaljournalofresearch.com

www.internationaljournalofresearch.com

ISSN no. **2348-6848**

ISBN no. **978-1-304-97715-1**

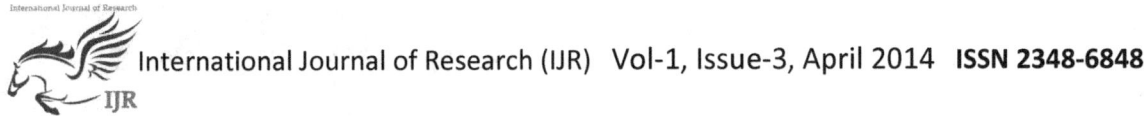

Board of Editors

We welcome editors of the scholarly articles and research papers. We will believe in sharing knowledge and helping fellow researchers and scholars in doing so. We as editors can pave the smooth way for the publication of quality papers in the open access International Journal of Research (IJR).

Member of the Board of Editors	Name	Details
Editor-in-Chief	S.N.Sharma	Indian Insitute of TechnologyRoorkee, India
Editor	Ashutosh Kumar Pandey	International School of Busines and MediaKolkata, India
Editor	Rajesh Kumar Prasad	National Institute of TechnologyRaipur, India
Editor	Harish Kumar	Nalanda Open UniversityPatna, India
Guest Editor	Arjun Satheesh	Indian Institute of Technology, Roorkee
Guest Editor	Sanil Kumar	Indian Institute of Technology, Roorkee

Either you have earlier edited papers or you wish to do so for the benefits of the larger community of the researchers and scholars, your presence and support will enhance our morale for serving the society through sharing of knowledge free of charge and making your expertise and experience known to the world.

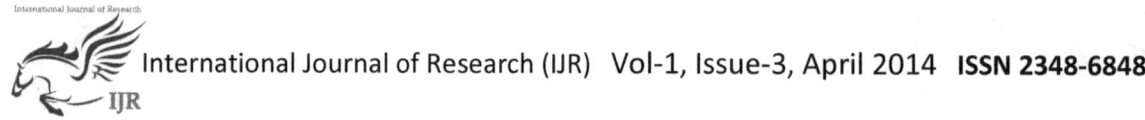

Forward

We are happy to announce the successful publication of the April Issue of the International Journal of Research. We encourage graduate, post graduate, PhD students and other scholar from the field of science, technology, planning and management to get their paper published and share the treasure of the knowledge that you have gained. To fulfil the aim and objectives of the Open Access Journal. We have made our content freely available for researchers and scholars to access and enhance their knowledge.

If you share this common view that scholarly content should be freely available and not locked up in the archives then it is time to come forward and be a part of the global mission of ensuring the publication and availability of the scholarly articles and research papers.

You can come forward to present the findings of the thesis or other research work in the form of a paper and make sure that your original thoughts and views reaches of the global audience. We have provided complete guidelines for the paper content-organization, formatting of the content etc. You can access them from our website.

The certificate of publication has been mailed to the authors through the email of the communing authors. We congratulate them and wish all the best for those who are going to send their article for publication in coming days.

You contact us through www.internationaljournalofresearch.com

And if your scholarly article or research paper is ready for publication then submit your manuscript through review@internationaljournalofresearch.com

-Regards

Editor-in-Chief

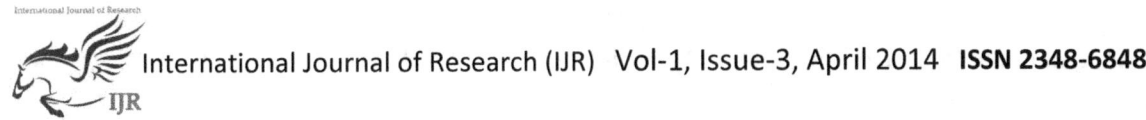

About the Journal

International Journal of Research, abbreviated as IJR is an international peer reviewed, internationally refereed, online, open-access journal published monthly. International Journal of Research (IJR) represents a revolution in scholarly journal publishing platform. A pioneering effort in liberal, open access publishing with fast and high quality peer review that brings journal publishing to the doorstep of every researcher and student. We believe that quality information should be free and accessible universally in this day and age. The ideology of an open-access journal is in being free for all and IJR will be free for all to read and share.

Publishing in the International Journal of Research

IJR takes special care to publish your research paper/article without any delay. Our journal aims to bring out the latent research talent and the professional work done by Scientists, Engineers, Architects, Planners, Practitioners, Administrators, Scholars, Graduate and Post Graduate students across all fields. This journal welcomes the submission of your research papers that meet our submission guidelines and the general criteria of significance and excellence in the field of Engineering, Science and Humanities. Submitted articles are peer reviewed by our panel of experts from various fields. All submitted papers are double-checked for plagiarized content. Please note that all submissions need to be previously unpublished.

http://internationaljournalofresearch.com/

You can mail your research papers/articles for publication to

submit@internationaljournalofresearch.com

You will be intimated after the review of the article if in case it merits publication in our journal. In case an article fails to meet our standards, but is found to be of scholarly merit, the author will be notified of possible modifications that would make the paper/article print-worthy.

Table of Content

Sr. No	Title of Paper or Article	Page No.
01	Publication Frequency and Contact Details	i
02	Board of Editors	ii
03	Forward	iii
04	About the Journal	iv
05	Sardar Hari Singh Nalwa – The Legend Hero of Punjab	01
06	Gurū- Śiṣhya paramparā: A strong education system of Indian Music	17
07	BEE (Bureau of energy efficiency) and GREEN BUILDINGS	24
08	To study the Emotional Intelligence of School Students of Haryana in Respect of Sex and Locale	34
09	An Approach to Healthy Life through Yoga in Ayurveda	42
10	Casteism and Women Empowerment: An Introspection	47
11	Synthesis and Characterization of Ni Doped Zno Nanoparticles	59
12	The Place of Customary International Law in the Nigerian Legal System – A Jurisprudential Perspective	67
13	OSU CASTE: A CRITIQUE	97

Sardar Hari Singh Nalwa - The Legend Hero of Punjab

SUREKHA[1]

Abstract

The present paper gives the information about the life of Sardar Hari Singh Nalwa. The paper is divided into three sections. First section deals with the early life of Hari Singh Nalwa. It gives information about his birth place, clan and his ancestors, it also described how Hari Singh meet Maharaja Ranjit Singh and why Maharaja adorned him a name of "Nall-wa". It also describes his governorship in Kashmir and Greater **Hazara. The second section deals with his conquest. Hari Singh Nalwa participated in the** twenty major battles and his military achievements in these battles were also described. The third section deals with his administration. He was sent to the most troublesome spots of the Sikh empire in order to "create a tradition of vigorous and efficient administration. Government of India in 2013 issued a postage stamp and marked the 176th anniversary of Hari Singh Nalwa's death. Besides this some other relevant information is also to mention here i.e. Hari Singh Nalwa's life became a popular theme for martial ballads. His earliest biographers were poets, including *Qadir Bakhsh urf Kadaryar, Misr Hari Chand urf Qadaryaar* and *Ram Dayal*, all in the 19th century. *Amar Chitra Katha* first published the biography of Hari Singh Nalwa in 1978. In 20th century, the song *Mere Desh ki Dharti* from the 1967 Bollywood film *Upkaar* eulogises him.

Keywords

Hari Singh Nalwa, Maharaja Ranjit Singh, military achievements, Sikh empire

Referring this Paper:

Surekha (2014). Sardar Hari Singh Nalwa - The Legend Hero of Punjab. *International Journal of Research (IJR)*. Vol-1, Issue-3.

---

[1] Surekha, Research Scholar, Dept of History, Panjab University, Chandigarh.

Sardar Hari Singh Nalwa was the great Commander-in-chief of the Sikh Army whose courage and strength are unparalleled in Sikh History. The indelible mark left by this son of Punjab in a short period of time during Maharaj Ranjeet Singh's reign will always remain a golden chapter in history. His administration and foresight were just a few of his qualities, which were responsible for his rise to number one in the Darbar of the Maharaja, and being promoted to serve as a governor of Kashmir and the volatile Peshawar. He was the only individual whose name was minted on the currency of Punjab. His name spelt terror into the hearts of the Afghans and the Afghan mothers used to silence their crying children by saying, "(quiet child), Khamosh bash- Haria raghle (Haria has come)!" Sardar Hari Singh was born in 1791 CE in Gujranwala (now in Pakistan) to Mata Dharam Kaur and father Sardar Gurdial Singh of the Uppal Khattri clan. S. Gurdial Singh was a commandant in the 'Shukerchakia Missal'. Hari Singh was only seven when his father died and it was in 1805 CE that he came under the attention of Maharaja Ranjeet Singh. In an open field event, which the Maharaja, used to organize. At an open field event, which the Maharaja used to regularly organize, Hari Singh showed his excellence in the events of horse riding, sword fighting, spear throwing and warfare etc., which completely astounded and pleased the Maharaja and immediately he invited Hari Singh to join his army. Hari's family was of 'Khatri' origin belonging to the Uppal tribe. They had migrated from Majitha, north of Amritsar, to Gujranwala in the eighteenth century. The first entry in the Nalwa family records maintained by the Pandas (Haridwar) was made in 1808 CE. It reads: "Kshattriya Uppal resident of Gujranwala, Hari Singh son of Gurdas Singhji and grandson of Bishen Singhji. Hari Singh's son is Gurdit Singhji.[2]

Hari Singh's grandfather was killed in 1762. His father accompanied Charrat Singh and Mahan Singh (Ranjit Singh's grandfather and father respectively) and on all their expeditions, and received in jagir the village of Balloke near Shahdera Hari lost his father at the age of seven.[3]

ਪੰਦਰਵੇਂ ਬਰਸ ਖੂਬ ਪਹਿਣਕੇ ਹਥਿਆਰ ਯਾਰੋ, ਪੈਦਲ ਘੋਲ ਕਰ ਸ਼ੇਰ ਬਾਘ ਸੁੱਟ ਮਾਰਦਾ। ਸੁੰਦਰ ਸਰੂਪ ਤੇਜ ਕੋਈ ਨ ਸੰਭਾਲ ਸਕੇ, ਮੁਖੜਾ ਦੀਦਾਰੀ ਜਿਵੇਂ ਲਾਲ ਦਮਕਾਰਦਾ। ਬਾਲਪਣ ਦੀ ਸੋਭਾ ਗੁਣ ਸੁਣ ਸੂਰਮੇ ਦੇ, ਛੋੜ ਦਿੱਤਾ ਸ਼ਤਰੂਆਂ ਨੇ ਅੰਨ ਪਾਣੀ ਘਾਰ ਦਾ। ਸੀਤਾਰਾਮਾ ਨਾਮ ਧਾਮ ਸੋਲਵੇਂ ਬਰਸ, ਹੋ ਗਿਆ ਮਸ਼ਹੂਰ ਹਰੀ ਸਿੰਘ ਸਰਦਾਰ ਦਾ॥੪॥

Pandrvein baras khoob pehen ke hathiyaar yaaro, paidal ghol kar sher bagh sutt maarda, Sundar saroop tej koi naa sambhaal sakke, mukhra deedari jivein laal damkaarda, Baalpan di shobha gunn sunn soorme de, chhod ditta shatruan ne ann-paani ghaarda, Sitarama naam dhaam solvein baras, ho gya mash-hoor Hari Singh Sardar da.[4]

Killing a lion

He was solely with the Maharajah for a number of months when someday he was asked to accompany the Maharajah for a hunt. As they entered the forest, suddenly a person feeding lion jumped on him and threw him on the ground. Hari Singh was completely caught unaware and failed to even have the opportunity to draw his weapon. However he got hold of the jaw of the lion and with great

---

[2] Vanit Nalwa, *Hari Singh Nalwa, "Champion of the Khalsaji" (1791-1837)*, Manohar Publishers, 2009, p. 9.

[3] L. Griffin, *the Panjab Chiefs*, Historical and Biographical Notices, T. C. Mc Cartney-Chronicle Press, Lahore, 1865, p. 184.

[4] Amarnath, *Zafarnama-i-Ranjit Singh (Persian)*, ed. Sita Ram Kohli, University of Panjab, Lahore, 1928, p. 31.

force flung the lion away, and withdrawing his weapon, with one blow cut off the lion's head. The Maharajah and therefore the different courtiers were extraordinarily stunned at this exploit. From that day forward Hari Singh was given the appellative of "Nalwa" by the Maharajah, who acknowledged that Hari Singh had killed the lion like the means King Nall wont to hunt (King Nall was a really brave king and was proverbial for his audacity to kill lions and different dangerous animals along with his vacant hands, and hence the maharajah adorned Hari Singh with that title i.e. "Nall-wa"- 'like Nall')[5]

Baron Hugel, a European traveller, writes in his book, Travels In Kashmir & the Punjab: "I surprised him by knowledge whence he had gained the appellation of Nalwa, and of his having cloven the head of a tiger, which had already seized him as its prey. He told the Diwan to bring some drawings and gave me his portrait, in act of killing the beast."

Governor of Kashmir 1820-21: Hari Singh Nalwa was appointed the first religion Governor of Kashmir in 1820. He ruled the province for somewhat over a year once the pull of the Sikh Forward Policy compelled his recall from the province. Hari Singh Nalwa was remembered in Kashmir for one thing he least expected. The currency minted while he was the governor had been the topic matter of much speculation.[6] Following his departure from this subah, all the coins minted below the Sikhs in this province were known as the *'Hari Singhee'*. Thereafter, no matter whom was the governor all coins minted in Kashmir continuing to be known as the *'Hari Singhee'* even following Hari Singh's death? Muslim and British historians criticised Hari Singh's tenure because the Governor of Kashmir. Deposit records show that their assessment was supported an incomplete understanding of the case.[7]

Jagirdar-Governor Greater Hazara 1822-37: The possibility of consolidating the North West Frontier of the Indian sub-continent into a province was presented by the relentless efforts of Sardar Hari Singh Nalwa. What he achieved during this region in a span of 15 years with limited resources and in the interior of a most turbulent population, was nothing in need of a miracle. Hazara, the crown of the Sindh Sagar Doab, was the most vital of all the territories below his Governance. His proceedings during this space present the best example of his talent as a military commander and as an administrator. The compiler of the Hazara lexicon acknowledged that Hari Singh Nalwa left his mark upon this district, that at that point solely a powerful hand like his may effectively control. "Of unbounded energy and courageousness, he was pitiless towards those who opposed his path. The town of Haripur befittingly perpetuates his name and therefore the fort of Harkishangarh forms an enduring monument of his power."[8]

It was on the 24th of February 1810, that the Maharaja, along with Sardar Hari Singh Nalwa, attacked Multan. This was a very hard battle where the Multan Fort was fortified by the Nawab of Bahawalpur, and even after considerable bombardment the walls of the fort held on. It was suggested that if some warriors could reach the fort and place

---

[5] Vanit Nalwa, *Hari Singh Nalwa, "Champion of the Khalsaji" (1791-1837)*, Manohar Publishers, New Delhi, 2009. p. 16.

[6] Surinder Singh, Coinage: Sovereignty to the Guru, in *Maharaja Ranjit Singh - Commemoration volume on Bicentenary of his Coronation 1801-2001* eds. Prithipal Singh Kapur and Dharam Singh, Punjabi University, Patiala, 2001, p. 81.

[7] Ganeshi Lal. *Siyahat-i-Kashmir (Kashmir Nama or Tarikh-i-Kashmir) by March-June 1846*, tr. Vidya Sagar Suri, 1955, Simla: The Punjab Government Record Office Pub. Monograph No. 4

[8] *NWFP Gazetteers - Gazetteer of the Hazara District 1907*, Chatto and Windus, London, 1908.

dynamite near the walls and blow the wall apart, the army then could enter the fort. This was a suicidal mission, but Sardar Hari Singh was the first volunteer to jump and except the challenge. He, along with 74 others did the needful and the Sikh army entered the fort and the battle was won, but Hari Singh was very seriously wounded, and had no hope of living. But after some time he recovered, to the delight of the Maharaja and the Sikh army, who now considered him an exceptional soldier, and was duly honoured by the Maharaja with more estate and money. His further conquests included Mitha Tiwana, Uch, and the historic win over the Afghanis at the Attock Fort. Later on 20th April 1819, the Sikh army under the command of Sardar Hari Singh Nalwa attacked Kashmir. A very ferocious battle was fought but eventually Kashmir became a part of the Sikh Empire. Diwan Moti Ram was given the governorship of Kashmir, but he proved a weak administrator, and was replaced by Sardar Hari Singh Nalwa on 24th August 1820. Sardar Hari Singh governed Kashmir in such an excellent manner that the Maharaja was highly pleased with him, and to reward him, the Maharaja instructed the Kashmir mint to name the currency after Hari Singh Nalwa. The "Hari Singh Rupee" can presently be seen in museums. The Maharaja needed Hari Singh for other campaigns, and as the situation of Kashmir was under control, he requisitioned Hari Singh back to Lahore, where plans were discussed to bring more territories under the Sikh Raj. While returning from Kashmir Sardar Hari Singh conquered Mangli on the way, which was another great win. He reached Lahore on 28th November 1821, and the Maharaja was extremely overjoyed to see him and learn of the triumph of Mangli. The Sikh army started their campaign with taking Mungher, Hazara and Hari Pur, which was named after Sardar Hari Singh. At the fort of Khairababad, the Sikh army under the command of Sardar Hari Singh only numbered 8000 and the Afghani army numbered nearly 150000. But the sheer bravery and audacity of the Sikh warriors was enough to overcome such a powerful force. Sir Alexander Barnes in his book "Barnes Travels - in Bukhara" narrates this Sikh victory as a milestone in history. On 16th of October 1831 Sardar Hari Singh was amongst the prominent Sardars, who along with Maharaja Ranjeet Singh met the British Governor General Lord Bentinck at the historic meeting of Ropar.[9]

The twenty major battles of Hari Singh Nalwa (either participated or were in command):

Battle of Kasur 1807: Hari Singh's first significant participation in a Sikh conquest on assuming charge of an independent contingent was in 1807, at the capture of Kasur. This place had long been a thorn in the side of Ranjit Singh's power because of its proximity to his capital city of Lahore. It was captured in the fourth attempt. This attack was led by Maharaja Ranjit Singh and Jodh Singh Ramgarhia. During the campaign the Sardar showed remarkable bravery and dexterity.[10] The Sardar was granted a *jagir* in recognition of his services.[11]

Battle of Sialkot 1808: Ranjit Singh nominated Hari Singh Nalwa to take Sialkot from its ruler Jiwan Singh. This was his first battle under an independent command. The two armies were engaged for a couple of days, eventually seventeen year old Hari Singh carried the day.[12]

Battle of Attock 1813: The fort of Attock was a major replenishment point for all armies crossing the Indus. In the early 19th

---

[9] Autar Singh Sandhu, *General Hari Singh Nalwa*, Cunningham Historical Society, Lahore, 1932, p. 97-98.
[10] Ibid, p. 5.
[11] Gulcharan Singh, *General Hari Singh Nalwa, the Sikh Review*, 1976, p. 36.
[12] Opcit, p. 8.

century, Afghan appointees of the Kingdom of Kabul held this fort, as they did most of the territory along this frontier. This battle was fought and won by the Sikhs on the banks of the Indus under the leadership of Dewan Mokham Chand, Maharaja Ranjit Singh's general, against Azim Khan and his brother Dost Mohammad Khan, on behalf of Shah Mahmud of Kabul. Besides Hari Singh Nalwa, Hukam Singh Attariwala, Shyamu Singh, Khalsa Fateh Singh Ahluwalia and Behmam Singh Malliawala actively participated in this battle. This was the first victory of the Sikhs over the Durranis and the Barakzais.[13] With the conquest of Attock, the adjoining regions of Hazara-i-Karlugh and Gandhgarh became tributary to the Sikhs. In 1815, Sherbaaz Khan of Gandhgarh challenged Hari Singh Nalwa's authority and was defeated.[14]

Abortive attempt on Kashmir 1814: The Sikhs attempted to take Kashmir soon after the Battle of Attock. The army was under the general command of Maharaja Ranjit Singh, who camped at Rajauri. The troops were led towards Srinagar by Ram Dayal, grandson of Dewan Mokham Chand, while Jamadar Khushal Singh commanded the van, Hari Singh Nalwa and Nihal Singh Attariwala brought up the rear. Lack of provisions, delay in the arrival of reinforcements, bad weather and treachery of the allies forced the Sikhs to retreat. The next few years were spent in subduing Muslim chiefs within the Kashmir territory, en route Srinagar Valley.[15] In 1815–16, Hari Singh Nalwa attacked and destroyed the stronghold of the traitorous Rajauri chief.[16]

Conquest of Mahmudkot (Mehmood Kot, Muzaffargarh 1816): In preparation of the conquest of the strongly fortified Mankera, Ranjit Singh decided to approach it from its southern extremity. After the Baisakhi of 1816, Misr Diwan Chand, Illahi Bakhsh, Fateh Singh Ahluwalia, Nihal Singh Attariwala and Hari Singh Nalwa accompanied by seven paltans and the topkhana went towards Mahmudkot. When news of its conquest arrived, it left the Maharaja so elated at the success of Sikh arms that he celebrated this victory with the firing of cannons. Two years later, on their way to Multan, the Sikhs captured the forts of Khangarh and Muzzaffargarh.[17]

Battle of Multan 1818: The winter of 1810 saw a jubilant Sikh army stationed near Multan in the Bari Doab. They were riding high on the success of having conquered the Chajj Doab. The possession of the city of Multan was taken with little resistance; however, the fort could not be captured. The fort was bombarded and mined without effect. Sardar Nihal Singh Attariwala and the young Hari Singh Nalwa were seriously wounded. A fire pot thrown from the walls of the fort fell on Hari Singh and he was so badly burnt that it was some months before he was fit for service.[18] Ranjit Singh was disconcerted beyond measure at the length of the siege and perforce had to abandon the attempt. Multan was finally conquered under the nominal command of Kharak Singh and the actual command of Misr Diwan Chand. It was a fiercely contested battle in which Muzzaffar Khan and his sons defended the place with exemplary courage, but they could not withstand the onslaught of

---

[13] G.S. Nayyar, *The Campaigns of General Hari Singh Nalwa,* Panjabi University Patiala, 1995, p. 89-90.
[14] Amar Singh, *Chamakda Hira Ya Jiwan Britant Sardar Hari Singh Nalwa,* Anglo-Sanskrit Press. Lahore, 1903. p. 112-13.
[15] Autar Singh Sandhu, *General Hari Singh Nalwa,* Cunningham Historical Society, Lahore, 1932, p.14.

[16] Ibid, p. 9.
[17] G.S. Nayyar, *The Campaigns of General Hari Singh Nalwa,* Panjabi University Patiala, 1995, p. 88.
[18] Autar Singh Sandhu, *General Hari Singh Nalwa,* Cunningham Historical Society, Lahore, 1932, p. 9.

the Sikhs. Hari Singh Nalwa was "chiefly instrumental" in the capture of the citadel.[19]

Peshawar becomes tributary 1818: When Shah Mahmud's son, Shah Kamran, killed their Barakzai Vazir Fateh Khan in August 1818 the Sikhs took advantage of the resulting confusion and their army formally forded the Indus and entered Peshawar, the summer capital of the Kingdom of Kabul (modern-day Afghanistan), for the first time. Thereafter, Hari Singh Nalwa was deputed towards Peshawar in order to keep the Sikh *dabdaba kayam* — maintain the pressure.

Mitha Tiwana 1818: In the beginning of 1819, Hari Singh accompanied Misr Diwan Chand to collect tribute from the Nawab of Mankera. On completion of the mission, Diwan Chand crossed the river Chenab along with his topkhana and set up his camp in Pindi Bhattian near Chiniot. He was asked to leave Hari Singh stationed in the suburbs of Nurpur and Mitha Tiwana. Hari Singh must have achieved significant success for soon thereafter the Maharaja bestowed all the possessions of the Tiwana chiefs in *jagir* on the Sardar.[20]

Kashmir becomes a part of the Punjab 1819: In April 1819, the Sikh army marched towards Kashmir. On this occasion, Prince Kharak Singh held nominal command. Diwan Chand led the vanguard, while Hari Singh Nalwa brought up the rear for the support of the leading troops. The third division, under the personal command of Maharaja Ranjit Singh, expedited supplies and conveyed these to the advance troops.[21] On the morning of 5 July 1819, the Sikh columns advanced to the sound of bugles. A severe engagement took place between the two armies and the Sikhs captured Kashmir. Great rejoicing followed in the Sikh camp and the cities of Lahore and Amritsar were illuminated for three successive nights. Thus came to an end the five centuries of Muslim rule in Kashmir.[22] Two years later, as Governor of Kashmir, Hari Singh Nalwa put down the rebellion of the most troublesome Khakha chief, Gulam Ali.[23]

Battle of Pakhli 1819: Under the Afghans, Hazara-i-Karlugh, Gandhgarh and Gakhar territory were governed from Attock. Kashmir collected the revenue from the upper regions of Pakhli, Damtaur and Darband. Numerous attempts by the Sikhs to collect revenue from Hazara-i-Karlugh not only met with failure, but also the loss of prominent Sikh administrators and commanders. Following the Sikh conquest of Kashmir, tribute was due from Pakhli, Damtaur, and Darband. On his return to the Punjab plains from the Kashmir Valley, Hari Singh and his companions followed the traditional *kafila* (caravan) route through Pakhli hoping to collect tribute from the region. The Sikh request for Nazrana resulted in the usual "fighting and mulcting"; the party however, was successful in their mission.[24]

Battle of Mangal 1821: Hari Singh's most spectacular success in the region of Pakistan's Hazara came two years later. On the successful conclusion of his governorship of Kashmir, he departed from the Valley and crossed the river Kishenganga at Muzaffarabad with 7000-foot soldiers. Hari Singh Nalwa traversed the hazardous mountainous terrain successfully, however when his entourage reached Mangal (Mangli, Pakistan) he found his passage opposed. Mangal, the ancient capital of Urasa was now the stronghold of the chief

---

[19] Gulcharan Singh, *General Hari Singh Nalwa, the Sikh Review*, 1976, p. 37.
[20] Opcit, p. 16.
[21] G.S. Nayyar, *The Campaigns of General Hari Singh Nalwa*, Panjabi University Patiala, 1995, p. 94.
[22] Autar Singh Sandhu, *General Hari Singh Nalwa*, Cunningham Historical Society, Lahore, 1935, p. 15.
[23] Ibid, p. 22.
[24] Ibid, p. 16.

of the Jaduns who controlled the entire region of Damtaur. Hari Singh requested the tribesmen for a passage through their territory, but they demanded a tax on all the Kashmir goods and treasure he was taking with him. All trade *kafilas* routinely paid this toll. Hari Singh's claim that the goods he carried were not for trade purposes was not accepted. When parleying produced no result, battle was the only option. A combined tribal force numbering no less than 25,000 gathered from all the adjoining areas and challenged Hari Singh and his men. Despite being completely outnumbered, the Sardar stormed their stockades and defeated his opponents with a loss to them of 2,000 men. Hari Singh then left to join forces with the Sikh army poised for an attack on Mankera, but after he had collected a fine from every house and built a fort in this vicinity.[25]

Battle of Mankera 1822: The Sindh Sagar Doab was chiefly controlled from Mankera and Mitha Tiwana. Nawab Hafiz Ahmed Khan, a relative of the Durranis, exerted considerable influence in this region. Besides Mankera, he commanded a vast area protected by 12 forts. With the weakening of Afghan rule in Kabul, the governors of Attock, Mankera, Mitha Tiwana and Khushab had declared their independence. Ranjit Singh celebrated the Dussehra of 1821 across the river Ravi, at Shahdera. Hari Singh, Governor of Kashmir, was most familiar with the territory that the Maharaja had now set his eyes on. Nalwa was summoned post-haste to join the Lahore Army already on its way towards the river Indus. The Maharaja and his army had crossed the Jhelum when Hari Singh Nalwa, accompanied by his Kashmir platoons, joined them at Mitha Tiwana. The Sikhs commenced offensive operations in early November. Nawab Hafiz Ahmed's predecessor, Nawab Mohammed Khan, had formed a cordon around Mankera with 12 forts—Haidrabad, Maujgarh, Fatehpur, Pipal, Darya Khan, Khanpur, Jhandawala, Kalor, Dulewala, Bhakkar, Dingana and Chaubara. The Sikh army occupied these forts and soon the only place that remained to be conquered was Mankera itself. A few years earlier, the Nawab of Mankera had actively participated in the reduction of Mitha Tiwana. The Tiwanas, now feudatories of Hari Singh Nalwa, were eager participants in returning that favour to the Nawab. The force was divided into three parts—one column being under Hari Singh—and each column entered the Mankera territory by a different route; capturing various places enroute all three columns rejoined near Mankera town. Mankera was besieged, with Nalwa's force being on the west of the fort.[26]

The fort of Mankera stood in the middle of the Thal. It was built of mud with a citadel of burnt brick surrounded by a dry ditch. To make the central fortress inaccessible, no wells were permitted by the Nawab to be sunk within a radius of 15 Kos. During the night of 26 November Hari Singh Nalwa, together with other chiefs and *jagirdars*, established their *morchas* (batteries) within long gunshot of the place. They found old wells, which their men cleared out and fresh ones were dug. On the nights of 6–7 December, they approached closer to the ditch. The ensuing skirmish was ferocious and resulted in considerable loss of life. The siege of the fort of Mankera lasted 25 days. Finally, the Nawab accepted defeat and the last Saddozai stronghold fell to the Sikhs. The Nawab was allowed to proceed towards Dera Ismail Khan, which was granted to him as *jagir*.[27] His descendants held the area until 1836.

Battle of Nowshera (Naushehra) 1823: The Sikhs forayed into Peshawar for the first time in 1818, but did not occupy the territory. They

---

[25] Ibid, pp. 25-26.

[26] Gulcharan Singh, *General Hari Singh Nalwa, the Sikh Review*, 1976, p. 38.

[27] Khushwant Singh, *Ranjit Singh Maharaja of the Punjab*, Penguin Books India, New Delhi, 2001, p. 138.

were content with collecting tribute from Yar Mohammed, its Barakzai governor. Azim Khan, Yar Mohammed's half-brother in Kabul, totally disapproved of the latter's deference to the Sikhs and decided to march down at the head of a large force to vindicate the honour of the Afghans. Azim Khan wanted to avenge both, the supplication of his Peshawar brethren and the loss of Kashmir. Hari Singh Nalwa was the first to cross the Indus at Attock to the Sikh post of Khairabad; he was accompanied by Diwan Kirpa Ram and Khalsa Sher Singh, the Maharaja's teenaged son, besides 8,000 men.

The Kabul Army was expected near Nowshera, on the banks of the river Kabul (Landai). Hari Singh's immediate plan was to capture the Yusafzai stronghold to the north of the Landai at Jehangira, and the Khattak territory to its south at Akora Khattak. The latter was taken without difficulty however Jehangira was a masonry fort with very strong towers and the Yusafzais offered tough resistance. Hari Singh entered the fort and established his *thana* there.[28] The remaining troops re-crossed the Landai River and returned to their base camp at Akora. Mohammed Azim Khan had encamped about ten miles north-west of Hari Singh's position, on the right bank of the Landai, facing the town of Nowshera, awaiting Ranjit Singh's approach. The Sikhs had scheduled two battles – one along either bank of the Landai.

After Hari Singh had successfully reduced the tribal strongholds on either side of the river, Ranjit Singh departed from the fort of Attock. He crossed the Landai River at a ford below Akora, and set up his camp near the fort of Jehangira. The famous army commander Akali Phula Singh and the no less renowned Gurkha commander BAL Bahadur, with their respective troops, accompanied the Maharaja. The Barakzais merely witnessed the main action from across the river. Hari Singh Nalwa's presence had prevented them from crossing the Landai.[29] Eventually, the inheritors of Ahmed Shah Abdali's legacy fled the scene in the direction of Jalalabad chased by Hari Singh Nalwa and his men to the very mouth of the Khyber Pass.

Battle of Sirikot 1824: Sirikot lay less than ten miles to the north-west of Haripur. This Mashwani village was strategically placed in a basin at the top of the northeast end of the Gandhgarh Range, which made its secure location a haven for the rebellious chiefs in the entire region. Hari Singh Nalwa went towards Sirikot before the rains of 1824. It was another six months before the attempt produced conclusive results. The Sardar almost lost his life in the course of this expedition. Ranjit Singh's military campaign for the winter of 1824 was scheduled towards Peshawar and Kabul. While stationed at Wazirabad, he received an *arzi* (written petition) from Sardar Hari Singh[30] informing him that he and his men were overwhelmingly outnumbered – one Sikh to ten Afghans. Ranjit Singh marched to [Rohtas], from there to [Rawalpindi] and via [Sarai Kala] reached Sirikot. The news of the approach of the Sikh army led to an instant dispersal of the insurgents.

The increasing success of the Sikh arms greatly disappointed the Yusafzai and other tribes inhabiting the trans-Indus region of Khyber Pakhtunkhwa. The Battle of Nowshera convinced them of their extreme vulnerability. Not only had the Kabul Barakzais let them down, but their subsequent application to the British for help had also met with little success.

Battle of Saidu 1827: The redeemer of the Yusafzais came in the form of one Sayyid Ahmad,[31] who despite being a 'Hindki' was

---

[28] Opcit, p. 39.
[29] Ibid, pp. 39-40.
[30] The Akhbars, Times, London, 11th March 1825.
[31] Autar Singh Sandhu, *General Hari Singh Nalwa*, Cunningham Historical Society, Lahore, 1935, p. 4.

accepted as a leader by them. Budh Singh Sandhanwalia, accompanied by 4,000 horsemen, was deputed towards Attock to assist in suppressing the Yusafzai rebellion. The Maharaja's brief required him to thereafter to proceed towards Peshawar and collect tribute from Yar Mohammed Khan Barakzai. Budh Singh first heard of the Sayyid after he had crossed the Indus and encamped near the fort of Khairabad. Ranjit Singh was still on the sickbed when the news of the Sayyid's arrival, at the head of a large force of the Yusafzai peasantry, reached him. The gallantry of the Yusafzai defence in the Battle of Nowshera was still vivid in his mind. On receiving this news, he immediately put into motion all the forces that he could muster and immediately dispatched them towards the frontier.

The Barakzais in Peshawar, though outwardly professing allegiance to the Sikhs, were in reality in league with the insurgents. The Sayyid marched from Peshawar in the direction of Nowshera. Sardar Budh Singh wrote to the Sayyid seeking for a clarification of his intention. The Sayyid haughtily replied that he would first take the fort of Attock and then engage Budh Singh in battle.

Hari Singh Nalwa stood guard at the fort of Attock with the intention of keeping the Sayyid and his men from crossing the river until reinforcements arrived from Lahore. News had reached the Sikhs that the jihadis accompanying the Sayyid numbered several thousand. The battle between the Sayyid and the Sikhs was fought on 14 Phagun (23 February) 1827. The action commenced at about ten in the morning. The Muslim war cry of *Allah hu Akbar*, or "God is the greatest", was answered by the Sikhs with *Bole so nihal, Sat Sri Akal*, or "they who affirm the name of God, the only immortal truth, will find fulfilment". Ironically, the opposing forces first professed the glory of the very same God Almighty, albeit in different languages, before they commenced slaughtering each other. The cannonade lasted about two hours. The Sikhs charged at their opponents, routed them, and continued a victorious pursuit for six miles, taking all their guns, swivels, camp equipage, etc. The number of killed was not mentioned, but blood was said to have flowed in torrents. The Sayyid sustained a complete defeat despite his vastly superior numbers. He was compelled to retreat to the Yusafzai Mountains. It was reported that 8,000 Sikhs had defended themselves against an enraged population of 150,000 Mohammedans.[32] A salute was fired, illumination was ordered by drumbeat in the city of Lahore in honour of the victory.

Peshawar 1834: The actual occupation of the great city of Peshawar and its ruinous fort, the Bala Hisar, by the Sikhs was quite a comedy and a total anti-climax. It was a reflection of Sardar Hari Singh Nalwa's formidable reputation in 'Pashtunistan'. Masson arrived in Peshawar just in time to see the Sikhs take control of the city. His eyewitness account reports that the Afghans simply fled the place and Hari Singh Nalwa occupied Peshawar without a battle.[33]

Dost Mohammad Khan flees 1835: Hari Singh Nalwa was the governor of Peshawar when Dost Mohammed personally came at the head of a large force to challenge the Sikhs. Following his victory against Shah Shuja at Kandahar, in the first quarter of 1835, Dost Mohammed declared himself *padshah* (king), gave a call for jihad and set off from Kabul to wrest Peshawar from the Sikhs. Ranjit Singh directed his generals to amuse the Afghans with negotiations and to win over Sultan Mohammed Khan. He directed them that on no account, even if attacked, were they to enter into a general engagement until his arrival.[34]

Hari Singh Nalwa and the other Sikh chieftains requested Ranjit Singh to permit

---

[32] Gulcharan Singh, *General Hari Singh Nalwa, the Sikh Review*, 1976, p. 40.
[33] Opcit, pp. 50-51.
[34] Opcit, p. 41.

them to engage with the Kabul Afghans. On 30 Baisakh (10 May 1835), Sardar Hari Singh, Raja Gulab Singh, Misr Sukh Raj, Sardar Attar Singh Sandhanwalia, Jamadar Khushal Singh, the Raja Kalan (Dhian Singh), Monsieur Court, Signor Avitabile, Sardar Tej Singh, Dhaunkal Singh, Illahi Bakhsh of the topkhana, Sardar Jawala Singh and Sardar Lehna Singh Majithia were ordered to move. The troops fanned out over five Kos, forming a semicircle in front of the Amir's encampment. Sardar Hari Singh proposed that the water of the stream Bara, which flowed in the direction of Dost Mohammed Khan's camp, be dammed. When the Ghazis appeared, Sardar Hari Singh commenced firing his guns. The Maharaja, however, prohibited him from indulging in battle and dispatched his Vakils to negotiate with the Amir.

Once Dost Mohammed Khan was assured that the Sikhs would affect a truce until their Vakils were in his camp, he let them know what he really felt. Harsh words were exchanged. He accused Fakir Aziz-ud-din of making "use of much language, having plenty of leaves but little fruit". On finding both his step brothers, Jabbar and Sultan, irredeemably lost to him, Dost Mohammed decided to retire from the field with the whole of his army, armament and equipage. He left at night, making sure that the Fakir did not return to the Sikh camp until after he had gone through the Khyber Pass.[35]

Jamrud (Khyber Pass) 1836: In October 1836, following the Dussehra celebrations in Amritsar, Hari Singh made a sudden attack on the village of Jamrud, at the mouth of the Khyber Pass. The Misha Khel Khyberis, the owners of this village, were renowned for their excellent marksmanship and total lack of respect for any authority. Hari Singh Nalwa's first encounter with this tribe had taken place following the Battle of Nowshera when he had pursued the fleeing Azim Khan; and once again, when he chased Dost Mohammed Khan in 1835.

The occupation of Jamrud was rather strongly contested, but it appeared that the place was taken by surprise. On its capture, Hari Singh Nalwa gave instructions to fortify the position without delay. A small existing fort was immediately put into repair. News of this event was immediately transmitted to Kabul. Masson informed Wade of the passage of events along this frontier in a letter dated 31 October 1836. With the conquest of Jamrud, at the very mouth of the Khyber,[36] the frontier of the Sikh Empire now bordered the foothills of the Hindu Kush Mountains

Panjtaar 1836: The defeat of the Khyber is sent shock waves through the Afghan community. However, more was to follow. Hari Singh Nalwa accompanied by Kanwar Sher Singh, now proceeded towards the Yusafzai strongholds, north-east of Peshawar, which had withheld tribute for three years. The Sikhs completely defeated the Yusafzais, with their chief, Fateh Khan of Panjtaar, losing his territory.[37] It was reported that 15,000 mulkia fled before the Sikhs like a herd of goats, many being killed and the remaining taking refuge in the hills. After burning and levelling Panjtaar to the ground, Hari Singh returned to Peshawar realising all the arrears of revenue. Fateh Khan was obliged to sign an agreement to pay tribute on which condition Panjtaar was released. When news of the conquest of Panjtar reached the Court of Lahore, a display of fireworks was proposed.

Battle of Jamrud 1837: The news of the conquest of Jamrud put Dost Mohammed Khan into a state of greatest alarm. General

---

[35] F.S. Waheeduddin, *The Real Ranjit Singh*, (second edition) Lion Art Press, Karachi, 2001, p. 73.

[36] Gulcharan Singh, *General Hari Singh Nalwa, the Sikh Review*, 1976, p. 41.

[37] G.S. Nayyar, *The Campaigns of General Hari Singh Nalwa*, Panjabi University Patiala, 1995, p. 152.

Hari Singh's latest possession gave the Sikhs the command of the entrance into the valley of Khyber. "If this was a prelude to further aggressive measures," the Amir "saw in the intimation and submission of the people of Khyber, the road laid open to Jalalabad." Were the Sikhs to take Jalalabad, their next stop would be Kabul. This information was followed by the intelligence of the defeat of the Panjtaris.

The Maharaja's grandson, Nau Nihal Singh was getting married in March 1837. Troops had been withdrawn from all over the Punjab to put up a show of strength for the British Commander-in-chief who was invited to the wedding. Yar Mohammed Khan has been invited to the great celebration. Hari Singh Nalwa too was supposed to be at Amritsar, but in reality was in Peshawar (some accounts say he was ill[38]) Dost Mohammed had ordered his army to march towards Jamrud together with five sons and his chief advisors with orders not to engage with the Sikhs, but more as a show of strength and try and wrest the forts of Shabqadar, Jamrud and Peshawar.[39] Hari Singh had also been instructed not to engage with the Afghans till reinforcements arrived from Lahore.

Hari Singh's lieutenant, Mahan Singh, was in the fortress of Jamrud with 600 men and limited supplies. Hari Singh was in the strong fort of Peshawar. He was forced to go to the rescue of his men who were surrounded from every side by the Afghan forces, without water in the small fortress. Though the Sikhs were totally outnumbered, the sudden arrival of Hari Singh Nalwa put the Afghans in total panic. In the melee, Hari Singh Nalwa was accidentally grievously wounded. Before he died, he told his lieutenant not to let the news of his death out till the arrival of reinforcements, which is what he did. While the Afghans knew that Hari Singh had been wounded, they waited for over a week doing nothing, till the news of his death was confirmed. By this time, the Lahore troops had arrived and they merely witness the Afghans fleeing back to Kabul.[40] Hari Singh Nalwa had not only defended Jamrud and Peshawar, but had prevented the Afghans from ravaging the entire north-west frontier. The Afghans achieved none of their stated objectives. The loss of Hari Singh Nalwa was irreparable and this Sikh victory was as costly as a defeat.[41]

Victories over the Afghans were a favourite topic of conversation for Ranjit Singh. He was to immortalise these by ordering a shawl from Kashmir at the record price of Rs5000, in which were depicted the scenes of the battles fought with them. Following the death of Hari Singh Nalwa, no further conquests were made in this direction. The Khyber Pass continued as the Sikh frontier till the annexation of the Punjab by the British.[42]

Administrator:

Hari Singh's administrative rule covered one-third of the Sikh Empire. He served as the Governor of Kashmir (1820–21), Greater Hazara (1822–1837) and was twice appointed the Governor of Peshawar (1834-35 & 1836- his death). In his private capacity, Hari Singh Nalwa was required to administer his vast *jagir* spread all over the kingdom.[43] He was sent to the most troublesome spots of the Sikh empire in order to "create a tradition of

---

[38] Khushwant Singh, *Ranjit Singh Maharaja of the Punjab*, Penguin Books India, New Delhi, 2001, p. 193.
[39] F.S. Waheeduddin, *The Real Ranjit Singh*, (second edition) Lion Art Press, Karachi, 2001, p. 74.
[40] Gulcharan Singh, *General Hari Singh Nalwa*, the Sikh Review, 1976, p. 45.
[41] Ibid, p. 46 and See also Khushwant Singh, Ranjit Singh Maharaja of the Punjab, Penguin Books India, New Delhi, 2001, p. 194.
[42] M.M. Mathews, *An Ever Present Danger*, Combat Studies Institute Press, Kansas, 2010, p. 15.
[43] Autar Singh Sandhu, *General Hari Singh Nalwa*, Cunningham Historical Society, Lahore, 1935, p. 123.

vigorous and efficient administration".[44] The territories under his jurisdiction later formed part of the British Districts of Peshawar, Hazara (Pakhli, Damtaur, Haripur, Darband, Gandhgarh, Dhund, Karral and Khanpur), Attock (Chhachch, Hassan Abdal), Jhelum (Pindi Gheb, Katas), Mianwali (Kachhi), Shahpur (Warcha, Mitha Tiwana and Nurpur), Dera Ismail Khan (Bannu, Tank, and Kundi), Rawalpindi (Rawalpindi, Kallar) and Gujranwala. In 1832, at the specific request of William Bentinck, the Maharajah proposed a fixed table of duties for the whole of his territories. Sardar Hari Singh Nalwa was one of the three men deputed to fix the duties from Attock (on the Indus) to Filor (on the Sutlej).

In Kashmir, however, Sikh rule was generally considered oppressive,[45] protected perhaps by the remoteness of Kashmir from the capital of the Sikh empire in Lahore. The Sikhs enacted a number of anti-Muslim laws,[46] which included handing out death sentences for cow slaughter,[47] closing down the Jamia Masjid in Srinagar, and banning the *azaan*, the public Muslim call to prayer.[48] Kashmir had also now begun to attract European visitors, several of whom wrote of the abject poverty of the vast Muslim peasantry and of the exorbitant taxes under the Sikhs.[49]

The Sikh rule in lands dominated for centuries by Muslims was an exception in the political history of the latter. To be ruled by *'kafirs'* was the worst kind of ignominy to befall a Muslim.[50] Before the Sikhs came to Kashmir (1819 CE), the Afghans had ruled it for 67 years. For the Muslims, Sikh rule was the darkest period of the history of the place, while for the Kashmiri Pandits (Hindus) nothing was worse than the Afghan rule.[51] The Sikh conquest of Kashmir was prompted by an appeal from its Hindu population. The oppressed Hindus had been subjected to forced conversions, their women raped, their temples desecrated, and cows slaughtered.[52] Efforts by the Sikhs to keep peace in far-flung regions pressed them to close mosques and ban the call to prayer because the Muslim clergy charged the population to frenzy with a call for 'jihad' at every pretext.[53] Cow-slaughter (Holy Cow) offended the religious sentiments of the Hindu population and therefore it met with severe punishment in the Sikh empire. In Peshawar, keeping in view "the turbulence of the lawless tribes ... and the geographical and political exigencies of the situation" Hari Singh's methods were most suitable.[54]

Nalwa was also a builder. At least 56 buildings were attributed to him, which included forts, ramparts, towers, gurudwaras, tanks, *samadhis*, temples, mosques, towns, *havelis*, *sarais* and gardens.[55] He built

---

[44] Kirpal Singh, *Historical Study of Maharaja Ranjit Singh's Times*, National Book Shop, Delhi, 1994, p. 98.

[45] T.N. Mada, "Kashmir, Kashmiris, Kashmiriyat: An Introductory Essay", in Rao, Aparna, *The Valley of Kashmir: The Making and Unmaking of a Composite Culture?* Manohar Publication, Delhi, 2008, p. 15.

[46] Chitralekha Zutshi, *Language of belonging: Islam, Regional Identity, and the making of Kashmir*, Oxford University Press, New York, 2003, p. 39.

[47] Victoria Schofield, *Kashmir in conflict: India, Pakistan and the Unending War*, I. B. Tauris, London, 2010, p. 5.

[48] Chitralekha Zutshi, Opcit, pp. 40-41.

[49] Victoria Schofield, Opcit, p. 6.

---

[50] Kirpal Singh, *Historical Study of Maharaja Ranjit Singh's Times*, National Book Shop, Delhi, 1994, p. 100.

[51] G. M. D. Sufi, *Kashmir Being a History of Kashmir From the Earliest Times to Our Own* (Reprinted ed.), Life and Light Publishers, New Delhi, p. 750.

[52] Autar Singh Sandhu, *General Hari Singh Nalwa*, Cunningham Historical Society, Lahore, 1935, p.13.

[53] Amar Singh, *Chamakda Hira Ya Jiwan Britant Sardar Hari Singh Nalwa*, Anglo-Sanskrit Press. Lahore, 1903. p. 115.

[54] Gulcharan Singh, *General Hari Singh Nalwa*, the *Sikh Review*, 1976, pp. 53-54.

[55] Krishan Lal Sachdeva, "Hari Singh Nalwa – A

the fortified town of Haripur in 1822. This was the first planned town in the region, with a superb water distribution system.[56] His very strong fort of Harkishengarh, situated in the valley at the foothill of mountains, had four gates. It was surrounded by a wall, four yards thick and 16 yards high. Nalwa's presence brought such a feeling of security to the region that when Hügel visited Haripur in 1835-6, he found the town humming with activity.[57] A large number of Khatris migrated there and established a flourishing trade. Haripur, tehsil and district, in Hazara, North-West Frontier Province, are named after him. Nalwa contributed to the prosperity of Gujranwala, which he was given as a *jagir* sometime after 1799,[58] which he held till his death in 1837.

He built all the main Sikh forts in the trans-Indus region of Khyber Pakhtunkhwa - Jehangira[59] and Nowshera on the left and right bank respectively of the river Kabul, Sumergarh (or Bala Hisar Fort in the city of Peshawar),[60] for the Sikh Kingdom. In addition, he laid the foundation for the fort of Fatehgarh, at Jamrud (Jamrud Fort).[61] He reinforced Akbar's Attock fort situated on the left bank of the river Indus[62] by building very high bastions at each of the gates. He also built the fort of Uri in Kashmir.[63]

A religious man, Nalwa built Gurdwara Panja Sahib in the town of Hassan Abdal, south-west of Haripur and north-west of Rawalpindi in Pakistan, to commemorate Guru Nanak's journey through that region.[64] He had donated the gold required to cover the dome of the Akal Takht within the Harmandir Sahib complex in Amritsar.[65] Following Hari Singh Nalwa's death, his sons Jawahir Singh Nalwa and Arjan Singh Nalwa[66] fought against the British to protect the sovereignty of the Kingdom of the Sikhs, with the former being noted for his defence in the Battle of Chillianwala.

For decades after his death, Yusufzai women would say "Raghe Hari Singh" ("Hari Singh is coming") to frighten their children into obedience.[67] A commemorative postage stamp was issued by the Government of India in 2013, marking the 176th anniversary of Nalwa's death. Hari Singh Nalwa died fighting the Pathan forces of Dost Mohammed Khan of Afghanistan. He was cremated in the Jamrud Fort built at the mouth of the Khyber Pass in Khyber Pakhtunkhwa. Babu Gajju Mall Kapur, a Hindu resident of Peshawar, commemorated his memory by building a memorial in the fort in 1892.[68]

Great Builder", in Kapur, P. S., Perspectives on Hari Singh Nalwa, ABS Publication, Jalandhar, 1993, p. 74.
[56] P.S. Kapur, & Khushwant Singh, "A Forward Base in the Tribal Areas", in Kapur, P. S.; Dharam, Singh, *Maharaja Ranjit Singh*, Patiala: Punjabi University, Patiala, 2001, p. 163.
[57] Kirpal Singh, *Historical Study of Maharaja Ranjit Singh's Times*, National Book Shop, Delhi, 1994, p. 99.
[58] Ibid, p. 87.
[59] Autar Singh Sandhu, *General Hari Singh Nalwa*, Cunningham Historical Society, Lahore, 1935, p. 24.
[60] S.M. Jaffar, *Peshawar: Past and Present*, S. Muhammad Sadiq Khan, Peshawar, 1945, p. 123.
[61] Ibid, p. 121.
[62] Kirpal Singh, *Historical Study of Maharaja Ranjit Singh's Times*, National Book Shop, Delhi, 1994, p. 102.

---

[63] G. M. D. Sufi, *Kashmir Being a History of Kashmir From the Earliest Times to Our Own* (Reprinted ed.), Life and Light Publishers, New Delhi, p. 729.
[64] Mohammad Waliullah Khan, *Sikh Shrines in West Pakistan*, Department of Archaeology Ministry of Education and Information, Government of Pakistan, Karachi, 1962, p. 17.
[65] Madanjit Kaur, *The Golden Temple: Past and Present* (Revised ed.), Guru Nanak Dev University, Amritsar, 2004, p. 214.
[66] Sohan Lal Suri identifies Jawahir Singh Nalwa and Arjan Singh Nalwa.
[67] Olaf, Caroe, *The Pathans 550BC-AD1957*, Macmillan and Co. Ltd, London, 1985, p. 313.
[68] Autar Singh Sandhu, *General Hari Singh Nalwa*, Cunningham Historical Society, Lahore, 1935, p. 85.

Hari Singh Nalwa's life became a popular theme for martial ballads. His earliest biographers were poets, including *Qadir Bakhsh urf Kadaryar*, *Misr Hari Chand urf Qadaryaar* and *Ram Dayal*, all in the 19th century. *Amar Chitra Katha* first published the biography of Hari Singh Nalwa in 1978. In 20th century, the song *Mere Desh ki Dharti* from the 1967 Bollywood film *Upkaar* eulogises him.

References:

1. Amarnath, *Zafarnama-i-Ranjit Singh (Persian),* ed. Sita Ram Kohli, University of Panjab, Lahore, 1928, p. 31.
2. Amar Singh, *Chamakda Hira Ya Jiwan Britant Sardar Hari Singh Nalwa,* Anglo-Sanskrit Press. Lahore, 1903. p. 115.
3. Autar Singh Sandhu, *General Hari Singh Nalwa,* Cunningham Historical Society, Lahore, 1935.
4. Chitralekha Zutshi, Language of belonging: Islam, Regional Identity, and the making of Kashmir, Oxford University Press, New York, 2003, p. 39.
5. F.S. Waheeduddin, *The Real Ranjit Singh,* (second edition) Lion Art Press, Karachi, 2001, p. 73-74.
6. Ganeshi Lal. *Siyahat-i-Kashmir (Kashmir Nama or Tarikh-i-Kashmir) by March-June 1846,* tr. Vidya Sagar Suri, 1955, Simla: The Punjab Government Record Office Pub. Monograph No. 4
7. G. M. D. Sufi, *Kashmir Being a History of Kashmir From the Earliest Times to Our Own* (Reprinted ed.), Life and Light Publishers, New Delhi, p. 729-750.
8. Griffin, *the Panjab Chiefs,* Historical and Biographical Notices, T. C. Mc Cartney-Chronicle Press, Lahore, 1865, p. 184.
9. G.S. Nayyar, *The Campaigns of General Hari Singh Nalwa,* Panjabi University Patiala, 1995, p. 152.
10. Gulcharan Singh, General Hari Singh Nalwa, the Sikh Review, 1976, pp. 40-54.
11. Kirpal Singh, *Historical Study of Maharaja Ranjit Singh's Times,* National Book Shop, Delhi, 1994, p. 99-105.
12. Krishan Lal Sachdeva, "Hari Singh Nalwa – A Great Builder", in Kapur, P. S., Perspectives on Hari Singh Nalwa, ABS Publication, Jalandhar, 1993, p. 74.
13. Khushwant Singh, Ranjit Singh Maharaja of the Punjab, Penguin Books India, New Delhi, 2001, p. 193-4.
14. Madanjit Kaur, *The Golden Temple: Past and Present* (Revised ed.), Guru Nanak Dev University, Amritsar, 2004, p. 214.
15. M.M. Mathews, *An Ever Present Danger*, Combat Studies Institute Press, Kansas, 2010, p. 15.
16. Mohammad Waliullah Khan, *Sikh Shrines in West Pakistan*, Department of Archaeology Ministry of Education and Information, Government of Pakistan, Karachi, 1962, p. 17.
17. *NWFP Gazetteers - Gazetteer of the Hazara District 1907*, Chatto and Windus, London, 1908.P.S. Kapur, & Khushwant Singh, "A Forward Base in the Tribal Areas", in Kapur, P. S.; Dharam, Singh, Maharaja Ranjit Singh, Patiala: Punjabi University, Patiala, 2001, p. 163.

18. S.M. Jaffar, *Peshawar: Past and Present*, S. Muhammad Sadiq Khan, Peshawar, 1945, p. 123.
19. Surinder Singh, Coinage: Sovereignty to the Guru, in *Maharaja Ranjit Singh - Commemoration volume on Bicentenary of his Coronation 1801-2001* eds. Prithipal Singh Kapur and Dharam Singh, Punjabi University, Patiala, 2001, p. 81.
20. The Akhbars, Times, London, 11th March 1825.
21. T.N. Mada, "Kashmir, Kashmiris, Kashmiriyat: An Introductory Essay", in Rao, Aparna, The Valley of Kashmir: The Making and Unmaking of a Composite Culture? Manohar Publication, Delhi, 2008, p. 15.
22. Vanit Nalwa, *Hari Singh Nalwa, "Champion of the Khalsaji" (1791-1837)*, Manohar Publishers, New Delhi, 2009. p. 16.
23. Victoria Schofield, Kashmir in conflict: India, Pakistan and the Unending War, I. B. Tauris, London, 2010.

Gurū- Śiṣhya paramparā: A strong education system of Indian Music

Ruchi Mishra

Abstract-

Ajnaān timirāṅdhasya jnaāṅjana Śalākyā
chkṣaūrūnmilitam yen tasmai shrī gurūve namaḥ

It means the Gurū is the one who can carry their disciples from the dark world of illiteracy (ajnaān) to the light of wisdom. The Gurū- Shishya parampara has been an inevitable part of education in the ancient Indian Culture. It said in our Veda's that Gurū hold the top most position in human's life the topmost place amongst all. In Indian Music tradition all knowledge is passing through oral tradition so Gurū is the only person who can suggest the right path to his/her disciple. Gurū-Śhishya Paramparā or the teacher – disciple relationship is an important part of India's teaching tradition. In ancient India most of the knowledge was passed on from the teacher to his pupil through oral tradition, this oral tradition of passing knowledge is known as the Gurū Śhishya Parampara. Indian classical music is still largely dependent on this tradition. In this paper I had tried to describe the details about this tradition.

Key words-

Gurū, Śhishya, Parampara, Indian Music

For Referring this Paper-

Misra, Ruchi (2014). Gurū- Śiṣhya paramparā: A strong education system of Indian Music. *International Journal of Research (IJR).* Vol-1, Issue-3.

Introduction-

When we are talking about Indian culture we can't ignore the importance of Art in it. The Indian culture is very rich because of their Art, Philosophy, and the spirituality. In Indian Philosophy the Gurū is explained as the God. The Gurū-Śhishya parampara has been an inevitable part of education in the ancient Indian Culture. In ancient India Gurū was the only source of knowledge. It is said that

Gurū Brhmā Gurū Vishnu Gurū Devo Maheshwarā
Gurū sakṣat parambrhmā tasmai shrī Gurūve namaḥ

It means Gurū is equal to the god like- Brhmā, Vishnū. In ancient time enyone who wanted to earn some knowledge about anything they should go to Gurukul for it. Gurukul was the place where the Gurū lived. The Śhishya had to live the same place for a fixed time like- 2-3 years. They have to do their scheduled work besides that they should have continued their study also because of this tradition of a living and learning relationship between the Gurū and the Student (Śiṣhya), signifying the emotional, intellectual and spiritual bonding between them. This strong bond between the Gurū and the Śhishya enables the Gurū to become a mentor who leads the Śiṣhya from ignorance to wisdom, and enlightenment.

Discussion-

In the world of Indian Music, the place of a Gurū has been considered as the highest of all so first of all we should know **the literal meaning of Gurū**. The word Gurū is made with two words like- 'Gu' and 'ru ' the word Gu is the symbol of darkness and ru is the symbol of light so Gurū is the person who leads the Śiṣhya from ignorance to wisdom, and enlightenment. Gurū is the one who is heavy with the weight of vast knowledge. Through devotion and meditation, the guru has experienced oneness with Divine Reality. Ideally such a being can awaken the divine within a student, transmit knowledge, offer guidance, and help integrate these experiences. The guru helps build the raft of knowledge which will cross over all evil and help to find the spiritual bliss. In Indian tradition the Gurū is also known as- Pram Gurū, Sata Gurū, Mahā Gurū, Shrī Gurū. It is said in Nātyaśāshtra that a **Gurū should have these qualities** like-

- *Smṛti- memory*
- *Mati - insight*
- *Medha- intelligence*
- *Uha- wisdom*
- *Apoha- willpower*

- *Śiṣya Nishpadana- production of good disciples.*

These are the general qualities of a Gurū described in **Nāṭyaśāsatra** but a Gurū should have some more qualities like-

- *Magnetic glow of an awakened personality*
- *Capacity to improvise songs and rhythmic sequences*
- *Flair for new creation in style*
- *Resourcefulness in handling situations*
- *Absolute mastery of technique integrating the body and soul of the dance art*
- *A live sense of rhythm and tempo in all their subtleties, expertise in conducting the dance ensembles*
- *Acquaintance with the individualities of musical instruments, sound knowledge of tradition acquired from seasoned veterans*
- *Perfect identification and devotion to ideals*
- *An intuitive perception of the strength and weakness of a student.*

The Gurū embodied divine power and was capable of bestowing grace. Through grace, the student, who may not yet have earned the merit, has love and favor freely transmitted to him or her from the Gurū now it is proved that how the place of Gurū was important to the knowledge and spiritual bliss.

As we know that Gurū Śiṣya parmpars is about to the relationship between the Gurū and the Śiṣhya so now we have to know some Characteristics about Śiṣya. The Śiṣya is the one who are the taker. In Urdu the Śiṣya is also called Śagird. Here the 'Śa' stands for the teacher and 'Gird' means around. Thus the word Śagird litlerary means one who makes the teacher the center of his or her world. It is a very sensitive and strong bond he/her is worshipping his/her Gurū and taking the knowledge and bliss. The Śiṣya should have these qualities as described in Nāṭyaśāsatra -

- *Smṛtī- memory*
- *Shlaghna- Merit*
- *Raga- devotion and dedication*
- *Saṅgharṣa- Great effort and hard work*
- *Medha- intelligence*
- *Utsaha- enthusiasm*

A student should have some more qualities like-

- *He should have belief in God, simple and pure life, possession of good*
- *He should win his Guru's confidence*
- *A student should learn with interest, practice his lessons sincerely, earn good name for his Guru and family*
- *He should adjust & adapt him to the various circumstances and situations*
- *He should spend most of the time with his guru and must have a thirst for more knowledge.*
- *He should be honest, self controlled, devoted to his work, cultured, disciplined, considerate, kind, helping and caring towards his class mates*
- *He should patient while learning, innovative, creative, update of new developments and inventions in the music forms*
- *He should proud of his Guru and devoted to his art form.*
- *He should not be short tempered, overconfident and boastful of himself and jealous of others*
- *He should set an ideal for the other disciples.*

The Gurū- Śiṣya parampara is a tradition where the student learnt the knowledge from his/her Gurū it's an oral tradition. It's an old education system.

The word of Education is made with a Sanskrit dhatu 'Śkṣha' it means to educate others. It is said in Manusmṛti that 'vidyāṃtamanuṣate' it means education makes man. The knowledge of Art's subject is also necessary with any other subjects. Education makes over inner thinking and Art's develop our thinking in aesthetics point of view and music is the topmost art form in amongst all art form and music education is very old tradition in Indian culture and the Gurū-Śiṣya parmpara is the way of music education.

Gurū Śiṣhya Parampara has been the most ancient and is also known as the best system. From the times of Veda, Music education has been given by the Gurū orally which is known as Gurū-mukh. The student while living in a Gurukul would offer services to the Gurū and at the same time, living under a stringent discipline, spending moderate lifestyle and perpetually practicing whatever education has been given to the student by the Gurū and learning by heart was the only way to receive knowledge. In ancient times music and other education depended upon the Guru. There was no syllabus and neither there was any provision to write the notations of the Bandishes learnt.

The student would try to imitate exactly the same way as rendered by the Guru. In the old days the practical aspect of music was given more weightage. The system did not give much importance to the intellectual treatment but the rules of the practical wing and primary values were not violated and only by the Gurū- Śiṣhya Parampara the student would graduate and become an artist.

Gurū- Śiṣhya parampara is based on the unwavering faith in the teacher and determination on the behalf of the disciple to make the mentor's path his own. This relationship is life long and deep. In the field of Indian classical dance and music this relationship is formalized with a ceremony called the Gandabandhan. With this ceremony the Gurū formally accepts the student in the lineage of the Gharana/ kul/ parampara the guru comes from Thus we see that though the term Gurū- Śiṣhya is considered Indian, it does have strong roots all the cultures and philosophies around the world. The bond between the mentor and his disciple is universal.

Conclusion-

In the Indian tradition place of a Gurū has been considered as the highest of all and the Gurū-Śiṣya parmpara is a one of the oldest tradition of Indian culture Even in today fast paced modern world we see this bond going strong not only in the field of dance and music but even in sports, science and academics. Everyone has some mentor in their life; it can be a parent, friend, a favorite teacher or coach, spouse or even some stranger who teaches you some valuable lesson in life. But for everyone irrespective to gender class and race our biggest teacher is our life itself. Music is a Gurū-Mukh vidya (Oral Tradition) so the Gurū is the only convener of this subject. In music the disciple is always following his/her guru to learn it. This involved the tradition of a living and learning relationship between the Gurū and the Student (Śiṣhya), signifying the emotional, intellectual and spiritual bonding between them.

References-

1. *Sangeet ShikshaAnk,* January-February 1988, Hanthrasa, Haridwar, India

2. Rishitosh, Dr. Kumar (2009). *Sangeet Shikshan keVividh Aayaam,* Kanishka Publishers, New Delhi, India.

3. Bharat, Muni (2003). Translator- *Shashtri Babulal, Natyshashtra, chaukhambha Sanskrit sansthan,* Varanasi

4. Altekar A.S (2001). *Education in Ancient India, Gyan Books, 2001, New Delhi, India.*

5. Awasthi, S.S (1988). *A Critique of hindusthani music and music education,* Dhanpat Rai and Son's, New Delhi, India.

6. C, Gangadhar (2010). *Theory and Practice of Hindu Music and the Vina tutor,* University of Michigan, USA.

7. Clements Sir Ernest, *Introduction to the study of Indian Music,* Nabu Press, New Delhi, India.

Other sources of reference-
- *http://en.wikipedia.org/wiki/Guru*
- *http://en.wikipedia.org/wiki/Guru-shishya_tradition*
- *http://www.hindupedia.com/en/Guru-Sishya_parampara*

Contact Information:

Ruchi Mishra
Research scholar, vocal department
Faculty of performing Arts
Banaras Hindu University
Email- ruchimishra379@gmail.com

BEE (Bureau of energy efficiency) and GREEN BUILDINGS

Dr. Supriya Vyas[69], Ar. Seemi Ahmed[70] and Ar. Arshi Parashar[71]

Abstract:

Green buildings are designed to reduce the overall impact of the built environment on human health and the natural environment. *This paper enumerates the various methods of bringing in energy efficiency. It introduces the Bureau of Energy Efficiency BEE (a national body) and presents an overview of the "Energy Conservation Building Code" (ECBC). In this paper, a case study of IIT* Kanpur has been *presented in detail. Despite its huge campus spread over an area of 4.3 square kilometres (1,100 acres) surprisingly, it uses only 605 kWh/m2 per year. Some more "Energy Efficient" buildings of India are also mentioned .In view of fast depleting energy reserves, energy conservation is need of the hour. Efficiently design homes and offices can cut energy bill substantially.*

The Bureau of Energy Efficiency, the relevant code and the Star Rating program, will go a long way to encourage energy efficiency. This program will help identify the careless owners who are frittering away the precious sources of energy. Continued support from the State Governments and the private sector is essential for the success of the program.

We need to wake up and act- and act fast before it is too late. Looking to the dwindling sources of energy and the danger of climatic deterioration caused by high carbon emissions, energy efficient buildings are necessity of today.

Keywords:

BEE (Bureau of Energy Efficiency), ECBC (Energy Conservation Building Code), GRIHA, Energy efficient buildings.

Referring this Paper

Vyas, Supriya; Ahmed, Seemi and Parashar, Arshi (2014). BEE (Bureau of Energy Efficiency) and Green Buildings. International Journal of Research (IJR). Vol-1, Issue-3. Pg-23-32

---

[69] Assistant Professor, Department of Architecture and Planning,
M.A. National Institute of Technology, Bhopal India,

[70] Assistant Professor, Department of Architecture and Planning,
M.A. National Institute of Technology, Bhopal India,

[71] Student, M.Plan (IV Sem), Department of Architecture and Planning,
M.A. National Institute of Technology, Bhopal India,

Introduction:

Buildings cause a number of environmental problems as a result of their construction, operation and maintenance. They consume a large amount of energy and resources, affect the quality of urban air and water and contribute to climate change. This is because buildings are designed as per building codes without any reference to any major environmental impacts over their entire life cycle.

With recent exponential increase in energy pricing, the formerly neglected or underestimated concept of energy conservation has swiftly assumed great significance and potential in cutting costs and promoting economic development, especially in a developing-country scenario. Energy efficiency in buildings can be achieved through a multipronged approach involving adoption of bioclimatic architectural principles responsive to the climate of the particular location; use of materials with low embodied energy; reduction of transportation energy; incorporation of efficient structural design; implementation of energy-efficient building systems; and effective utilization of renewable energy sources to power the building. India is quite a challenge in this sense. Several buildings have come up, fully or partially adopting the above approach to design. Reckless and unrestrained urbanization, with The Building Energy Rating Certificate (BER) is part of the Energy Performance of Buildings EU Directive. The aim of the Directive, which first came into force in Ireland on 4 January 2003, is to make the energy performance of a building transparent and available to potential purchasers or tenants. Then it became a practice in India also, from 27th may 2007. The BER is simply a check to see how good your house or building is at using energy and will measure how much energy and carbon your building may typically use or produce over a given year. It is only concerned with the fabric of the dwelling and does not take account of occupant behaviour.

GREEN BUILDING:

Green building (also known as **green construction** or **sustainable building**) refers to a structure and using process that is environmentally responsible and resource-efficient throughout a building's life-cycle: from siting to design, construction, operation, maintenance, renovation, and demolition. This requires close cooperation of the design team, the architects, the engineers, and the client at all project stages. The Green Building practice expands and complements the classical building design concerns of economy, utility, durability, and comfort.

Although new technologies are constantly being developed to complement current practices in creating greener structures, the common objective is that green buildings are designed to reduce the overall impact of the built environment on human health and the natural environment by:

- Efficiently using energy, water, and other resources.
- Protecting occupant health and improving employee productivity.
- Reducing waste, pollution and environmental degradation.

Figure 1 below shows the benefits of Green building.

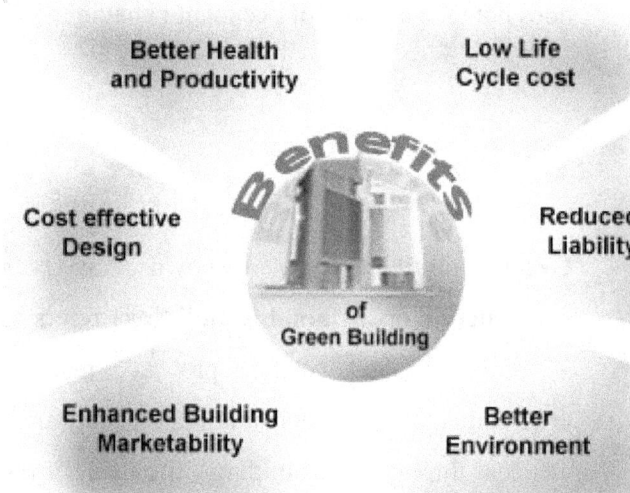

Fig 1: Green building benefits

BUREAU OF ENERGY EFFICIENCY (BEE):

Bureau of Energy Efficiency (BEE) is a statutory energy conservation body under the power ministry. The Government of India set up BEE on 1st March 2002 under the provisions of the Energy Conservation Act, 2001.

The Major Regulatory Functions of BEE include:

- Develop minimum energy performance standards and labeling design for equipment and appliances
- Develop specific Energy Conservation Building Codes
- Activities focusing on designated consumers
- Develop specific energy consumption norms
- Certify Energy Managers and Energy Auditors
- Accredit Energy Auditors
- Define the manner and periodicity of mandatory energy audits
- Develop reporting formats on energy consumption and action taken on the recommendations of the energy auditors

The Major Promotional Functions of BEE include:

- Create awareness and disseminate information on energy efficiency and conservation
- Arrange and organize training of personnel and specialists in the techniques for efficient use of energy and its conservation
- Strengthen consultancy services in the field of energy conservation
- Promote research and development
- Develop testing and certification procedures and promote testing facilities
- Formulate and facilitate implementation of pilot projects and demonstration projects
- Promote use of energy efficient processes, equipment, devices and systems

- Take steps to encourage preferential treatment for use of energy efficient equipment or appliances
- Promote innovative financing of energy efficiency projects
- Give financial assistance to institutions for promoting efficient use of energy and its conservation
- Prepare educational curriculum on efficient use of energy and its conservation
- Implement international co-operation programs relating to efficient use of energy and its conservation.

ENERGY CONSERVATION BUILDING CODES (ECBC):-

The Indian Bureau of Energy Efficiency (BEE) had launched the Energy Conservation Building Code (ECBC) on February 2007. The code is set for energy efficiency standards for design and construction with any building of minimum conditioned area of 1000 Sq. mts and a connected demand of power of 500 KW or 600 KVA. The energy performance index of the code is set from 90 kW·h/sq./year to 200 kW·h/sqm/year where any buildings that fall under the index can be termed as **"ECBC Compliant Building"** More over the BEE had launched a 5 star rating scheme for office buildings operated only in the day time in 3 climatic zones, composite, hot & dry, warm & humid on 25 February 2009.

Energy conservation building codes are mechanism to prescribe energy use/conservation in commercial buildings. These codes are formed to comply with energy consumption norms and standards and to prepare and implement schemes for its efficient use and conservation state governments have power to amend building codes to suits local and regional climatic conditions.

Energy conservation building codes set the minimum energy efficiency standards for design and construction at the same time in encourage energy efficient design without constrain on the building function comfort health or the productivity of the occupants with proper regard for economic considerations.

GRIHA- the National rating System

TERI, being deeply committed to every aspect of sustainable development, took upon itself the responsibility of acting as a driving force to popularize green building by developing a tool for measuring and rating a building's environmental performance in the context of India's varied climate and building practices.

The rating system called 'Green Rating for Integrated Habitat Assessment' (GRIHA) quantifies parameters like energy consumption, waste generation, renewable energy adoption over the entire lifecycle of the building. In 2007, it was adapted and adopted by the Ministry of New and Renewable Energy (MNRE) as the national rating system for green buildings in order to bring down the ecological impact of buildings in India to a nationwide acceptable level. GRIHA currently operates under ADARSH (Association for Development and Research on Sustainable Habitats) and is supported by the National Advisory Council (NAC) and Technical Advisory Committee (TAC).

It takes into account the provisions of the National Building Code (NBC) 2005, the Energy Conservation Building Code (ECBC) 2007 announced by BEE and other IS codes, local bye-laws, other local standards and laws. The system, by its qualitative and quantitative assessment criteria, would be able to 'rate' a building on the degree of its 'greenness'. The rating would be applied to new and existing building stock of varied functions – commercial, institutional, and residential.

A CASE STUDY: CESE Building, IIT KANPUR

Architect	Kanvinde Rai and Chowdhury Architects and Planners
Energy consultant	TERI (The Energy and Resources Institute)
HVAC consultant	Gupta Consultants and Associates
Electrical consultant	Kanwar Krishen Associates Pvt. Ltd
Landscape Architect	Mr. Yogesh Kapoor

ABOUT CESE:

The CESE building in IIT Kanpur became the first GRIHA rated building in the country and it scored 5 stars, highest in GRIHA under the system. It has become a model for green buildings in the country. It has proved that with little extra investment, tremendous energy and water savings are possible. There are various projects which are the first of their kinds to attempt for green building ratings like apartment residential buildings and non-air conditioned buildings. Measures are being taken to spread awareness about the GRIHA-National Green Building Rating System of India.

The CESE is a research facility at the IIT (Indian Institute of Technology), Kanpur on a plot area of 175 000 square metre (approximately 4.5 acres) (Fig 2) the facility houses laboratories, seminar rooms, and discussion rooms. Given the function of the building, it was decided that it should be designed in an environment friendly manner. The evaluation committee has awarded a final score of 93 out of 100 to the building. The building has incorporated many green features following the TERI-GRIHA recommendations. Some special features of this building are as follows:

-The building is fully compliant with the ECBC (Energy Conservation Building Code).
-Sustainable site planning has been integrated to maintain favorable microclimate.
-The architectural design has been optimized as per climate and sun path analysis.
-The building has energy-efficient artificial lighting design and daylight integration.
-It also has energy-efficient air conditioning design with controls integrated to reduce annual energy consumption.
-Passive strategies such as an earth air tunnel have been incorporated in the HVAC design to reduce the cooling load.

Figure 2: View of CESE building.

Performance

The EPI (Energy Performance Index) of the building is predicted to be 45.43 kWh/m2/annum, which is 41.3% less than the TERI GRIHA benchmark. In comparison to a conventional building, 59% energy savings are predicted in the CESE building. The Centre has attempted to conserve and utilize resources efficiently; recycle, reuse, and recharge the systems at every stage of design and construction.

Key Sustainable Features:

The building attempted various GRIHA criteria to make it into a green building. Few such criteria's are elaborated as follows:

Sustainable site planning

In order to minimize impact of site development on the environment and surroundings, several best practice guidelines were adopted like demarcation of site for construction, installation of dust screen around the disturbed area to prevent air pollution and spillage to undisturbed site area. Top soil was excavated, stored and preserved outside the disturbed construction site. Erosion control systems were adopted and several trees on site were protected. To increase the perviousness of site and to reduce heat island effect caused due to hard paving around the building, total paving around the building was restricted to 17%, and more than 50% of the paving is either pervious or shaded by trees. Irrigation water demand has been reduced by more than 50% in comparison to GRIHA benchmark. Adequate health and safety measures related to construction were taken.

Energy Conservation

-Renewable energy from photovoltaic panels provide annual energy requirements equivalent to 30% of internal lighting connected load.

- Hot water demand is met by solar hot water system

Figure 3. 1.2 KWp High Efficiency Solar Concentrator with Tracker being installed at IIT Kanpur Courtesy: Moser Baer Photovoltaic Pvt Ltd.

Architectural design:

Architectural design optimized as per the climate of Kanpur, sun path analysis, predominant wind direction, and existing vegetation.

There is a large water body to cool the micro climate.(Fig 4) The Orientation of building is North – South. There is daylight integration in all living spaces. . Common circulation areas are natural day lit and naturally ventilated through integration of skylights and ventilators.

Fig 4: Greenery and water body around the building

Efficient window design:

By selecting efficient glazing, external shading to reduce solar heat gain but at the same time achieve glare free natural daylight inside all the laboratory spaces of the building.

Portland Pozzolona Cement (PPC) with fly-ash content is used in plaster and masonry mortar.

Wood for doors is procured from commercially managed forests. Modular furniture made from particle board is used for interiors.

Landscape protection:

Existing trees have been preserved and protected. Roof shaded by bamboo trellis and green cover to reduce external solar heat gains from the roof.

Water cooled chiller has been selected that complies with the efficiency recommendations by the ECBC (Energy Conservation Building code).

Variable Frequency Drive installed in the Air Handling Units (AHUs).It also has energy-efficient air conditioning design with controls integrated to reduce annual energy consumption.

Earth air tunnel system is used for free cooling/heating of the building for a major part of the year. This technology uses the heat sink property of earth to maintain comfortable temperatures inside the building.

Figure 6. Roof shaded by bamboo trellis and green cover been incorporated in the HVAC design to reduce

Low energy strategies such as replacement of water cooler by water body to cool the condenser water loop, integration of thermal energy storage and earth air tunnels enabled reduction in chiller capacity.

Optimized architectural design and integration of energy efficient fixtures has resulted in the reduction in annual energy consumption by 41% from GRIHA's benchmark.

Water conservation
There are two ways of conserving water during post construction and after the building is occupied.

One is landscape water demand and second is building water demand. In this building, reduction in landscape water demand by more than 50% was achieved by use of minimum grass/lawn area, maximum green area under native vegetation and native trees. Low flow plumbing fixtures are used in the building resulting in reduced water consumption from GRIHA's benchmark in this building by 62%. Waste water is treated and reused for irrigation. Rain water harvesting has been designed. The building's water body adopts rainwater harvesting norms.

Figure 7. Maintained water bodies

Energy-efficient HVAC:
Depending on the brief and nature of the project, non-air-conditioned (non-AC) and air-conditioned (AC) spaces are treated differently. Non-AC areas are designed to maximize the thermal comfort levels with the use of natural ventilation, passive techniques, and low energy consuming evaporative cooling strategies. The AC areas on the other hand are designed to minimize the load of installed HVAC systems

and hence reduce energy consumption in building.

- Efficient fixtures, efficient lamps, Daylight integration have been used **in lighting system.**

CONCLUSION:

Looking to the dwindling sources of energy and the danger of climatic deterioration caused by high carbon emissions, energy efficient buildings are the necessity of today. Construction of energy efficient building like CESE in IIT, Kanpur should be encouraged more and more. It has become a model for green buildings in the country. It has proved that with little extra investment, tremendous energy and water, savings are possible.

The Bureau of Energy Efficiency, the relevant code and the Star Rating program, will go a long way to encourage energy efficiency. This program will help identify the careless owners who are frittering away the precious sources of energy. Continued support from the State Governments and the private sector is essential for the success of the program.

REFERENCES:

1. Proceedings of AMTID 2011, NIT Calicut, Jun 22-24, 2011 Ratings of Energy Efficient Buildings: A Case of IIT, Kanpur, by Dr. Supriya Vyas.
2. Majumdar, Mili; *Tata Energy Research Institute, India Ministry of Non-Conventional Energy Sources (2001)*. Energy-efficient Buildings in India. TERI Press. ISBN 9788185419824.
3. Chen, Olivia (2008-11-05). "Bamboo-Veiled Dormitory by Architecture BRIO"(http://www.inhabitat.com/ 2008/ 11/ 05/magic-bus-dormitory/).
4. U.S. Environmental Protection Agency. (October 28, 2009). Green Building Basic Information. Retrieved December 10, 2009, from http:/ /www. epa. gov/ greenbuilding/ pubs/ about. htm
5. Hopkins, R. 2002. *A Natural Way of Building.* (http:/ / transitionculture. org/ articles/ a-natural-way-of-building-2002/) Transition Culture. Retrieved: 2007-03-30.
6. Sharma, Shashikant Nishant (2014), Urban Forms in Planning and Design, *International Journal of Research (IJR)*, Volume-1, Issue-1.
7. Mili Majumdar, *Energy-efficient Buildings in India,* Tata Energy Research Institute, India, Ministry of Non-Conventional Energy Sources
8. *Book : Representative designs of energy-efficient buildings in India by TERI*
9. *http://www.nrg-builder.com/greenbld.htm*
10. *http://www.thefreelibrary.com/Green+architecture+in+India:+combining+modern+technology+with+...-a01695971*
11. "The Druk White Lotus School" (http:/ / www. pbs. org/ e2/ teachers/ teacher_207. html). PBS. . Retrieved 2009-02-05.
12. http:/ / www. cleantechfinland. fi/ solutions/ green construction/
13. http:/ / www. Publications. Parliament. uk/ pa/ cm200405/ cmselect/ cmenvaud/ 135/ 13507. htm#a23
14. *Environmental Performance Index 2010 (epi.yale.edu/)*

Contact Information:

Dr. Supriya Vyas
Assistant Professor, Department of Architecture and Planning,
M.A. National Institute of Technology, Bhopal India,
supriya_vyas@hotmail.com

Ar. Seemi Ahmed
Assistant Professor, Department of Architecture and Planning,
M.A. National Institute of Technology, Bhopal India,
Email: seemiahmed2002@yahoo.com

Ar. Arshi Parashar
Student, M.Plan (IV Sem), Department of Architecture and Planning,
M.A. National Institute of Technology, Bhopal India,
Email: arshiparashar@gmail.com

To study the Emotional Intelligence of School Students of Haryana in Respect of Sex and Locale

Dr. Archana Nara[72]

Abstract –

New concepts such as emotional intelligence have become more widely understood; more educators are realizing that cognitive ability is not the sole or critical determinant of young people's aptitude to flourish in today's society. Proficiency in emotional management, conflict resolution, communication and interpersonal skills is essential for children to develop inner self-security and become able to effectively deal with the pressure and obstacles that will inevitably arise in their lives. Objectives of this study were to study:

(i) To study the emotional intelligence of school students.
(ii) To compare emotional intelligence of male and female school students;
(iii) To compare emotional intelligence of rural and urban school students.

Sample of 800 secondary school students from four districts of Haryana were selected through random sampling method. Emotional Intelligence Inventory by Dr. S. K. Mangal and Mrs. Shubhra Mangal. The data was analysed by Product Moment Correlation and Z-test. The main findings are: (1) a significant difference was found in emotional intelligence of male and female school students. It is in favour of female students; (2) a significant difference was found in emotional intelligence of rural and urban school students. It is in favour of urban students.

Keywords-

Emotional Intelligence, School Students, Emotional Management, Conflict Resolution

For Referring this Paper:

Nara, Archana (2014). To study the emotional intelligence of school students of Haryana in respect of sex and locale. *International Journal of Research (IJR)*. Vol-1, Issue-3, Pg-33-39.

---

[72] Assist. Professor
C.R. College of Education
Delhi Road, Rohtak-124001,*Haryana, India*
E-mail: ansiwach@gmail.com

Introduction

To understand the concept of emotional intelligence it is important to have some clarity about the two terms that constitute it, namely, intelligence and emotion. Emotions are present in every activity of human being. They are prime movers of thought and conduct. They play important role in influencing physical health, mental health, social life, character, learning process and area of adjustment. When our feelings become intense and excited, they become emotion. What do happiness, fear, anger, affection, shame, disgust, surprise, lust, sadness, elation of love have in common? These are emotions which directly affect one's day to day life for long, it is believed that success at the work place depends on Intelligence Quotient (IQ) as reflected by one's academic achievements. But how bright is one outside the classroom, facing the life's difficult moments? Here we need a different kind of resourcefulness termed as emotional intelligence (EQ). How a college drop-out like Bill Gates managed to build such a vast empire, could be attributed to emotional intelligence. Mother Teresa who decided to devote her life as a nun to social service with no resource of her own, could successfully arouse world conscience to help the needy and the poor. Emotional intelligence gives person a competitive edge.

The term "Emotional Intelligence" was first used in an article in 1990 by Peter Salovey and John D. Mayer. They defined emotional intelligence as a type of social intelligence that involves the ability to monitor one's own and other's emotions, to discriminate among them, and to use the information to guide one's thinking and actions.

In emotional facilitation of thinking, emotions can be useful in terms of directing attention to pressing concerns and signalling what should be the focus of attention (George and Brief, 1996). Emotions can also be used in choosing among options and making decisions, being able to anticipate how one would feel if certain events took place can help decision makers choose among multiple options (Damasio, 1994). Emotions can be used to facilitate certain kinds of cognitive processes. For example, positive moods can facilitate creativity, integrative thinking and inductive reasoning, and negative moods can facilitate attention to detail, detection of errors and problems, and careful information processing (Salovey et al, 1993). Shifts in emotions can lead to more flexible planning, the generations of multiple alternative and a broadened perspective on problems (Salovey and Mayer, 1989). When people are in positive moods, they tend to be more optimistic and perceive that positive events are more likely and negative events are less likely whereas, when people are in negative moods they tend to be more pessimistic and perceive that negative events are more likely than the positive events. (Salovey and Birnbaum, 1989).

OBJECTIVES OF THE STUDY

1. To study the emotional intelligence of school students.
2. To compare emotional intelligence of male and female school students.
3. To compare emotional intelligence of rural and urban school students.

DESIGN AND METHODOLOGY

The sample for this study consisted of 800 students from secondary schools of four districts of Haryana (Rohtak, Sonepat, Hisar and Ambala). For sample selection stratified random sampling technique was used. Firstly, selections of schools were made on the basis of sex i.e. male and female schools; secondly selections of schools were made on the basis of locale i.e. urban and rural schools. Students studying in Xth class were taken from the above mentioned districts randomly. They constitute the population of this study.

Emotional Intelligence Inventory by Dr. S. K. Mangal and Mrs. Shubhra Mangal which designed for use with Hindi and English knowing of school, college and University students for the measurement of their emotional intelligence (total as well as separately) in respect of four areas or aspects of emotional intelligence namely, Intra–personal Awareness (Knowing about one's own emotions) Inter-personal Awareness (Knowing about others emotions), Intra-personal Management (Managing one's own emotions) and inter-personal Management (Managing others emotions) respectively.

STATISTICAL TECHNIQUES USED

Pearson Product Moment correlation 'r' was employed for determining the relationship between Home environment and Emotional intelligence.

Mean, Standard Deviation and 'Z' test was used to find out the significance of difference between the mean scores.

ANALYSIS AND INTERPRETATION:

OBJECTIVE NO. 1 To study the emotional intelligence of school students.

Table 1: Emotional Intelligence Scores of School Students

Emotional Intelligence Scores	School Students	%age
31-47	46	5.75
48-61	182	22.75
62-75	406	50.75
76-88	164	20.50
89-100	2	0.25
Total	800	100.00

Fig. 1: Emotional Intelligence Scores of School Students

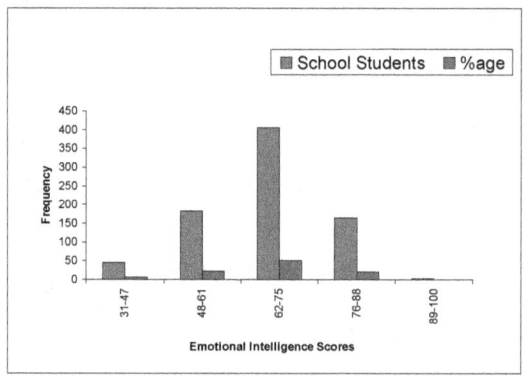

From Table 1, it is evident that only 5.75 percent school students' emotional intelligence scores lie in the range of 31 to 47 and 22.75 percent school students' emotional intelligence scores fall in the range of 48 to 61, 50.75 percent school students' emotional intelligence scores lie in the range of 62-65, 20.75 percent school students' emotional intelligence scores fall in the range of 76-88 and only 0.25 percent students' emotional intelligence scores lie in the range of 89 to 100.

So from the results, it is concluded that 50.75 percent students falling in the range of 62 to 75 are average in emotional intelligence.

Table 2: Distribution showing population mean and variability of emotional intelligence of school students.

N	$\overline{X}$	σ	σ_M
800	66.46	10.27	0.36

Result: 65.7 to 67.22 at 0.05 level of confidence

65.53 to 67.39 at 0.01 level of confidence

Hence with respect to our above data, there are 95 chances out of 100 that M_{pop} would fall between the score limits 65.7 to 67.22 and there are 99 chances out of 100 that the M_{pop} would fall between 65.52 to 67.39. Our confidence that these intervals contain M_{pop} is 95 percent or P of .95 & 99 percent or P of .99 respectively. It means that there are 5 percent chances that mean of population (M_{pop}) of emotional intelligence of school students would fall beyond the range 65.7 to 67.22 and there are 1 percent chances that M_{pop} of emotional intelligence would fall beyond the range of 65.53 to 67.39.

OBJECTIVE NO.2 To compare emotional intelligence of male and female school students

HYPOTHESIS There will be no significant difference in emotional intelligence of male and female school students

Emotional Intelligence Scores	Male	Female
31-40	10	1
41-50	45	13
51-60	73	70
61-70	158	171
71-80	90	108
81-90	24	36
91-100	0	1
Total	400	400

Fig. 2: Emotional Intelligence scores of male and female school students

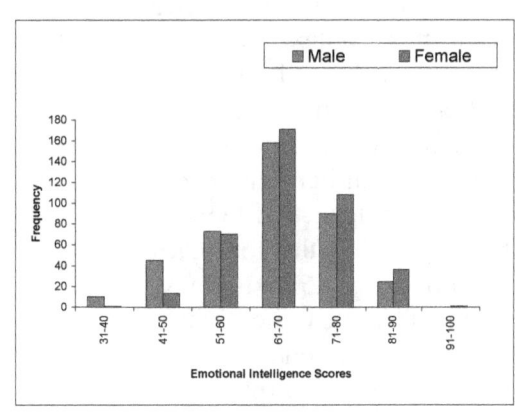

Table 3 Emotional Intelligence scores of male and female school students

Table 4: Genderwise $\overline{X}$, σ, N, σ_D and Z-value of Emotional Intelligence of male and female school students.

Gender	$\overline{X}$	□	N	□_D	Z-Value
Male	64.53	11.038	400	0.59	6.45*
Female	68.39	9.038	400		

*Significant at 0.01 level of significance

From Table 4, it is evident that the 'z' value of emotional-intelligence of male and female school students is 6.454 which is significant at 0.01 level of significance with df 798. It indicates that the mean scores of emotional intelligence of male and female school students differ significantly. In this context, the null hypothesis that "there will be no significant difference in emotional intelligence of male and female school students" is rejected. Further, the mean scores of emotional intelligence of females is 68.39 which is higher than that for males whose mean score is 64.53. It may, therefore, be concluded that the female school students are emotionally more intelligent than their male counterparts.

OBJECTIVE NO. 3 To compare emotional intelligence of rural and urban school students.

Hypothesis

There will be no significant difference in emotional intelligence of rural and urban school students.

Table 5: Emotional Intelligence scores of rural and urban school students

Emotional Intelligence Scores	Rural	Urban
31-40	8	3
41-50	21	37
51-60	85	58
61-70	181	148
71-80	75	123
81-90	30	30
91-100	0	1
Total	400	400

Fig. 3 Emotional Intelligence scores of rural and urban school students

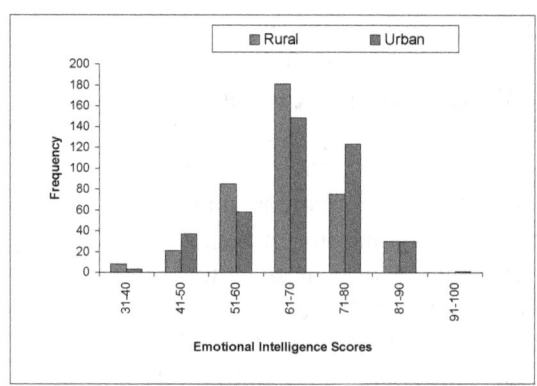

Table 6: Localitywise $\overline{X}$ □, N, □_D and Z-value of Emotional Intelligence of Rural and Urban School students

Locality	$\overline{X}$	□	N	□_D	Z-Value
Rural	65.77	9.89	400	0.58	2.36*
Urban	67.14	10.59	400		

*Significant at 0.05 level of 0.01 level of significance

From Table 6, it is evident that the 'z' value of emotional intelligence of rural and urban school students is 2.36 which is significant at 0.05 level of significance with 798 degree of freedom. It indicates that the mean score of emotional intelligence of rural and urban school students differed at 0.05 level of significance. In this context, the null hypothesis that "there will be no significant difference in emotional intelligence of rural and urban school students" is rejected. Further, the mean score of emotional intelligence of urban students is 67.14 which is slightly higher than that of rural students whose mean score is 65.77. It may, therefore, be concluded that urban students are emotionally more intelligent than rural school students.

Major Findings

1. A significant difference was found in emotional intelligence of male and female school students. It is in favour of female students

2. A significant difference was found in emotional intelligence of rural and urban school students. It is in favour of urban students.

significantly in rural and urban school students indicating that urban students are better in emotional intelligence than their rural counterparts.

Conclusion

Careful analysis and interpretation of data has revealed following conclusions:

The findings show that male and female students differ significantly on emotional intelligence. Therefore, it may be concluded that female students are emotionally more intelligent than male students. Consequently, female students belonged to home environment more conducive to emotional intelligence. This helps them in managing emotions better than male school students. The findings of the present study lead to the conclusion that emotional intelligence differs

References

1. Chopra Vanita. "Educational Implications of Emotional Intelligence for Better Teacher and Student Performance." *MERI Journal of Education*, 4(1), 2009.

2. Ciarrochi, J., Chan, A., Y.C. and Bajgar, J. "Measuring emotional intelligence in adolescents." *Personality and Individual Difference*, **31**(7), 2001, 1105-1119.

3. Devi, Uma and Rayules, T.R. "Levels of emotional intelligence of adolescent boys and girls", Journal *of Indian Psychology*, **32**(2), 2005.

4. Devi, Uma and Royal, U.T.R. "Adolescent's Perception about Family Environment and Emotional Intelligence", *Indian Psychological Review*, **62**(3), 2004, 157-67.

5. Dey, Niradhar. "The influence of emotional intelligence on academic self-efficacy and achievement", *Psycholingua.* **39**(2), 2009.

6. Finegan, J.E. "Measuring emotional intelligence: Where we are today." ERIC Service No. ED426087, 1998.

7. Goleman, D. *Working with Emotional Intelligence.* NY: Bantam Books, 1998.

8. Gakhar, S.C. and Manhas, K.D. "Cognitive correlates of emotional intelligence of adolescents." *Ram-Eash Journal of Education*, **2**(2), 2005, 78-83.

9. Graves, J. "Emotional intelligence and Cognitive Ability: Predicting Performance in Job Simulated Activities", *California School of Professional Psychology*, San Diego, 1999.

10. Jadhav and Patil. "Emotional intelligence among student teachers in relation to general intelligence and academic achievement." *Edu Track*, **10**(3), 2010, 36-37.

11. Mathur, Dube and Malhotra "Emotional intelligence interrelationships of attribution, taking responsibility and scholastic performance in adolescents", *Indian Review*, **60**(4), 2003, 175-180.

12. Narain Shruti and Vijayalakshmi. Emotional Intelligence and Academic Achievement of School Children, *Psycho-Lingua*, **40** (1&2), 2010, 80-83.

13. Neil Humphrey; Andrew Curran; Elisateth Morris; Peter Farrell; Devin Woods. "Emotional Intelligence and Education: A Critical Review." *EDUTRACK*, 2007.

14. Parker, J.D., Summerfeldt, L.O.J., Hogan, M.J., & Majeski, S.A. "Emotional intelligence and academic success: Examining the transition from high school to university." *Personality and Individual Differences*, **36**, 2004, 163-172.

15. Petrides, K.V., Frederickson, N., & Furnham, A. "The role of trait emotional intelligence in academic performance and deviant behaviour at school." *Personality and Individual Differences*, **36**, 2004, 277-293.

16. Singh, M. Chaudhary, O.P. and Asthana, M. Impact of locale and gender on emotional intelligence of adolescents. *Psycho-Lingua* ISSN: 0377-3132), 2008

17. Subramanyam, K. and Sreenivaas Rao K. Academic achievement and emotional intelligence of secondary school children, *Journal of Community Guidance and Research*, **25**, 2008.

18. Van Rooy, D., Alonso, A., & Viswesvaran, C. Group differences in emotional intelligence scores: Theoretical and practical implications. *Personality and Individual Differences*, **38**, 2005, 689-700.
19. Zeinder, M., Roberts, R.D. and Mathews, G. "Can emotional intelligence be schooled? A Critical review." *Educational Psychologist*, **37**(5), 215-231.

An Approach to Healthy Life through Yoga in Ayurveda
Dr. Devanand Upadhyay [73]

Abstract:

Yoga is the spiritual science for holistic development of physical, mental and spiritual aspect of living being. Ayurveda believes an interrelationship between psyche and body and thus if psyche is effected leads to an adverse effect on body and vice versa. Ayurveda is a science of living being which has its broad aim of living healthy life and curing of ailments. The instability of inner psyche (manas) is controlled through yoga. Bhagwat geeta emphasizes yoga as the state of *sama sthiti or* equilibrium in actions. Acharya Patanjali's *yoga darshan* has an impact on Ayurveda and later *hatha yoga pradeepika* and *gheranda samhita*'s various *yogasana, kriya, mudra, bandha and pranayama* were used as a part of treatment. The yogic concept of health and disease enables us to understand that the cause of physical disorder is result of higher levels of psyche (manas).Ayurveda believes that pragya paradh is root cause to diseases and thus pragya shodhan is very important. Ayurveda emphasis yoga and is a way to achieve atma gyana through pragya shodhana. Ayurveda has given definition of yoga, powers of yogi, scope of yoga, its implication in attainment to highest stage of moksha. Ayurveda and yogic methods can be applied for preventive, promotive and rehabilitatory health of human being.

Keywords:

Yoga, Ayurveda, health, *pragya*etc.

For Referring this Paper:

Upadhyay, Devanand (2014). An Approach to Healthy Life through Yoga in Ayurveda. *International Journal of Research (IJR)*. Vol-1, Issue-3. Pg-41-44

---

[73] BAMS, MD (Ay.)IMS,BHU
Lecturer(Samhita Siddhanta),VYDSAM, Khurja

dev.asdbhu10@gmail.com

Introduction:

The term yoga can be derived from either of two roots, *yujir yoga* (to yoke) or *yuj samādhau* (upanishada) meaning to unite, to combine or to integrate of individual to supreme. Yoga has also been popularly defined as "union with the divine" in other contexts and traditions. According to *srimad bhagwat geeta Yoga* is defined as

"yogah karmasu kaushalam" i.e. yoga is excellence in performance.

"samatvam yogamuchyate" (Bh.geeta 2/47) balanced state in situations like sukha dukha, and other situations.

"tam vidyat dukha sanyogaviyogam yogam sangyitam" (Bh.geeta 6/23) detachment of miseries is yoga.

The members of four divisions of human life namely *brahmachari, grihastha, vanaprastha, and sanyasi* are all meant to be yogis. Human life is not just for sense gratification. brahmachari under the care of spiritual master control the mind by abstaining from sense gratification. Similarly households can perform acts with great restraint (*Bh.gee.4/26*).

According to yoga sutra of **patanjali yoga** is defined as "the stilling of the changing states of the mind".[3] it has been recommended in 8 stages which are yama (restraints), niyama (observances), asana (physical postures), pranayama (breathing control), pratyahara (withdrawl of sense organs), dharnana (contemplation), dhyana (meditation), Samadhi (attainment of super consciousness).

Chronological history of yogic science: Yoga in ancient time was for attainment of moksha. Acharya charak has also given yoga as a way to attain moksha but with time this has been modified. Pre–philosophical speculations and diverse ascetic practices of first millennium BC were systematized into a formal philosophy in early centuries by the *Yoga Sutras of Patanjali*. By the turn of the first millennium, hatha yoga emerged from tantra. Along with its many modern variations, is the style that many people associate with the word *yoga* today. Gurus from India later introduced yoga to the west, following the success of Swami Vivekananda in the late 19th and early 20th century. In the 1980s, yoga became popular as a system of physical exercise across the Western world. This form of yoga is often called Hatha yoga. Many studies have tried to determine the effectiveness of yoga as a complementary intervention for cancer, schizophrenia, asthma, and heart disease.[7,8] Now a days yoga is more popular as health promoter, prevention from diseases loke diabetes mellitus, hypertension, bronchial asthma, obesity, stress condition of life.

Ayurvedic view towards yoga: *"Sukhartha sarva bhutanam matah sarva pravrittayah gyanagyana visheshattu margamarga pravrittayah" (ch.su.28/35)*

Action of every human being is to attain sukha but it depends on knowedge one has. One who follows path after proper knowledge is on verge to attain sukha while other attain dukha.

It has been said by acharya charaka that sites of vedana are *manas* (psyche), *deha* (body) and *indriyas* (senses). Yoga and moksha are considered to be two stages where there is complete eradication from all vedana. *Moksha* is said to be stage where there is complete eradication from vedana and yoga has been considered as a tool to achieve *moksha*. Stage of perception of *sukha* or *dukha* is attained from *sannikarsh* (union) of *atma, manas, indriya* and *indriyartha* as this union leads to any sort of knowledge and it is either perceived as *sukha or dukha*. *Sukha* is stage of *anukula vedana* (perception of good) that is achieved running on path of *dharma* while *dukha* is vice versa. When *manas* is said to be still and stable in self then due to non-attachment with any *karya*, it is free from feeling of *sukha* and *dukha*, and with *sharira*.

Process : Describing the process of atma gyana it has been established that

I[st] step- *Indriyas* should control from *artha* (objects).

IInd step- *Manas* should be controlled or mobility of manas should be controlled.

IIIrd step- *Manas* should attach introvert with *Atman* and it should realize *gyana of Atman*.

Ways to achieve:
1. *Sadavritta* (describes maximum yama and niyama) leading to indriya jaya.ch.su.
2. *achara rasayana* (various conducts having effect like rasayana or rejuvenator. Ch.ch.1/4
3. controlling of *kayika*(stealing objects, coitus with females of others ,violence),*vachika*(rough and rude language, abuse, untimely talk, complaining about others), *manasika dharaniya vega*(anger, greed, ignorance, shameless, ego, lust, envy)etc. Ch.su.7/28,29,30
4. follow paths described for *moksha marga* leading to *suddha manas* and *satya buddhi*.ch.sh.1,5

Comparison of pure manas: On following the methods given in *moksha marga* one gets free from *rajas* and *tamas* and purified manas looks similar to a mirror cleaned with oil and cloth or broom. It has been compared with sun which shines gracefully in absence of eclipse, cloud, storm, fog etc. the manas becomes still in self and it is glorified as *deep* inside a lantern.ch.sh.5/18

Qualities of satya buddhi*:* After attaining *suddha manas* a person attains *suddha buddhi* which is so powerful to break walls of moha or tamas. Human knows all the bhavas of production and becomes *nihaspriha*. After attaining *satya buddhi yoga siddhi* and *tattva gyana* occurs. This *satya buddhi* is *vidya,siddhi, mati, medha, pragya, gyana*. Ch. Sh. 5/19. It has been termed as *sthita pra gya* in geeta.

Powers of yogi: yogis attains 8 siddhi which are *aavesh*(can enter into body of others), *chetas gyana*(knows whats going in others mind), *arthanam chhandatah kriya*(can understand indriyartha through manas), *drishti*(can see which is not perceivable from eyes), *shrotra*(can heat non perceivable from ear), *smriti* (memory), *kanti*(lusture), *ishta darshna*(can disappear and appear acordng to will).ch.sh.1/141

Yoga and health:

Acording to swami kuvalayananda founder of kaivalya dhama, positive health does not mean freedom from disease but it is jubilant and energetic view of living and feeling that is peak lstate of living being at all levels – physical, mental, emotional, spiritual, and social. Ayurveda can be studied by all for the attainment of virtues, wealth and pleasure.[4] Virtues are attained by treating individuals who have spiritual knowledge, who practice and propagate righteousness and other like mother, father, brother, friend and superiors. These are also achieved by meditation, propagation and practice of the spiritual knowledge contained in the science of life. This has been called as para dharma i.e. best among all codes of right path. All these constitute the higher virtue of his life. The person endowed with psychic strength and sattvika character tolerates all by his own will power. Due to unregulated manas the action or performance of human runs on wrong way which produces diseases. Due to lack of smriti a human does not remember what is favourable or what is unfavourable for him. Due to abnormal state of dhi, dhriti and smriti human become unable to control the three types of voluntary actions like kayika, vacika manasika. *pragyaparadh* is the root cause of diseases, this pragyaparadh can be prevented through yoga. *Atma bala* developed as result of yoga has super capability to provide a clear concept of decisive actions. Adhyatma gyana is the only way of *pragya shodhan* so health can be maintained through yoga.

Discussion:

In early century yoga was considered as spiritual disciple practiced for moksha but with time it is now practiced as an alternative and contemporary medicine all around the world.it has built psychic stability in form of patience, positive attitude towards life, fitness in body, increased vital capacity. Yoga is physical, mental, and spiritual practices or disciplines which originated in ancient India with a view to attain a state of permanent peace of mind in order to experience one's true self. Early Charak has proposed *sadavritaa* which are code of conducts, *achara rasayana, dharaniya manasika vega* like desire, anger, greed, hatred, mosha marga etc. following these leads to conquer over senses and manas. Ones sattva is increased leading to diminution of *rajas and tamas* (psychological ailments) leads to stable *pragya* or intellect. This stable intellect not only thinks positively but also lead to personality improvement. Thus yoga is practical way to attain well-being and is advised to be practiced for good quality of life. Presently we are living a life full of stress. This stress brings disturbances in physical and psychic dosha s leading to diseases. The attitude to positive thinking comes from yoga. Yoga is free of cost, is safe and medication free. So it can be practiced by all leading to knowledge, purity of mind and intellect and healthy living being.

Conclusion:

Ayurveda has described yoga in plenty and various conducts to lead to yoga which leads to highest goal of salvation.

Abbreviations used: *ch.sh.* Charaka sharira sthana, *ch.su.* charaka sutra sthana

References:

1. Dasgupta, Surendranath (1975). *A History of Indian Philosophy* **1**. Delhi, India: Motilal Banarsidass. p. 226. ISBN 81-208-0412-0.
2. Srimad Bhagwat Geeta By A.C Bhakhtivedant Swamy, Bhativedant Book Trust Publication, Mumbai 2006
3. Sahu divya, Singh Rani, Optimism through yoga; Annual proceedings of recent advances in yoga, p.36, ISBN 978-81-925546-0-0.
4. Murthy KHHVSS Narsimha, Kumar Dileep: Non pharmacological approach in management of manas rogas through yoga and Ayurveda, pg.25 Annual proceedings of Recent advances in yoga, p.36, ISBN 978-81-925546-0-0.
5. Charak samhita with Vidyotini Hindi Commentary, Vol-I and II, shastri K.N. and Chaturvedi, G.N. (Ed. Shastri R.D. et al) Chaukhambha. Bharati Academy Varanasi 1984.
6. Sushruta Samhita of Maharsi Sushruta with Ayurvedatatwasandipika, Hindi Commentary, Shastri, a.d., Chaukhambha Sanskrit Sansthan, Varanasi, 1953-1959.
7. Smith, Kelly B.; Caroline F. Pukall (May 2009). "An evidence-based review of yoga as a complementary intervention for patients with cancer". *Psycho-Oncology* 18(5):465–475. doi:10.1002/pon.1411.PMID 18821529.
8. Vancampfort, D.; Vansteeland, K.; Scheewe, T.; Probst, M.; Knapen, J.; De Herdt, A.; De Hert, M. (July 2012). "Yoga in schizophrenia: a systematic review of randomised controlled trials". *Acta Psychiatrica Scandinavica* 126 (1): 12–20., art.nr. 10.1111/j.1600-0447.2012.01865.x
9. Sharma, Manoj; Taj Haider (Oct 2012). "Yoga as an Alternative and Complementary Treatment for Asthma: A Systematic Review". *Journal of Evidence-Based Complementary & Alternative Medicine* 17 (3): 212–217.doi:10.1177/2156587212453727.
10. Innes, Kim E.; Cheryl Bourguignon (November–December 2005). "Risk Indices Associated with the Insulin Resistance Syndrome, Cardiovascular Disease, and Possible Protection with Yoga: A Systematic Review". *Journal of the American Board of Family Medicine* 18 (6): 491–519.

Casteism and Women Empowerment: An Introspection

Rajib Bhattacharyya†††††††††††††††††††† & Tamal Gupta‡‡‡‡‡‡‡‡‡‡‡‡‡‡‡‡‡‡

Abstract:

Women are the integral part of the society. Society can't be an ideal society without the contribution made by the women. In the matter of active women empowerment in the society the contribution of law deserves a special remark at national and international level.

The pervasive inequalities existing in the societies are the basic gender issues that lead to deprivation of human rights for woman. Without human rights they can't have security of life and liberty as well as a dignified existence and they neither can realize their full potential as human beings nor can function as full citizens, participating in all the processes that contribute towards the social progress of a country. **Besides all the feminine legislations, international law and covenants and though judiciary has also played an active role with the spirit of judicial activism; still, the euphoria to justice for woman is seems to be a vague discourse without social transformation. Social transformation seeks for active participation of both the sex to move forward towards the ideal of woman empowerment and gender justice in their true sense.**

Therefore, in this paper various dimensions relating to women has been analyzed at the various stages and steps and also provided certain remedial or suggestive measures to improve the conditions of women which will ultimately lead to the betterment of the conditions of women in the society and then only the women will be empowered in the true sense of the term.

Keywords:

Social Change, Feminist Empowerment, Gender Biasness, Human Rights, Education, Liberty, Equality, Parliamentary Democracy, Women Empowerment, Revolutionary Approach, Family Jurisprudence etc.

For Referring this Paper:

Bhattacharya, Rajib & Gupta Tamil (2014). Casteism and Women Empowerment: An Introspection. International Journal of Research (IJR). Vol-1, Issue-3. Page 45-54

---

†††††††††††††††††††† (B.A-LL.B, LL.M, CCL, DEM, DHR, PGDBO), Assistant Professor, University Law College, Gawhati University, Guwahati- 14, Assam, India

‡‡‡‡‡‡‡‡‡‡‡‡‡‡‡‡‡‡ (B.A-LL.B, LL.M, DHR, DCL, CCL), Assistant Professor, The Oxford College of Law, Bangalore, Karnataka, India

Introduction:

Law is considered as the most pivotal instrument of social change through its dynamic approach. The law is also considered as the golden key to uplift the social status of women because women are playing the most cardinal role in every society whether it is developed or developing. The introduction of the Constitution of India created a revolution in the field of feminist empowerment. According to Manu "Gods were pleased in those houses where the women held in honour". If we turn the pages of Hindu mythology then we can see that women are called as "Shakti" which means "Power". Women are playing the various roles like mother, daughter, wife, sister and each and every time she is the inspiration of every successful man. They maintain the equilibrium of the civilized society and if there is any disregard of women then entire civilized society will collapse like a pack of cards. Therefore, according to Pt. Jawaharlal Nehru it can be stated that "you can tell the condition of a nation by looking at the status of its women". In regard of empowerment of women in the Indian society the Constitution of India deserves a special mention. The Constitution of India not only protects the women but also empowers the women through its various provisions as well as upholds the true spirit of feminist jurisprudence. In the present days woman don't want to confine within the four walls of the house and they also want to create separate identity in the society beside men. Now the gender biasness has been eliminated by virtue of the various legal instruments and there is a remarkable recognition of women contribution towards establishing an ideal state. The basic human rights like right to education, right to life and liberty, right to equality, right to take part in the parliamentary democracy etc. help them to create a special identity in the society from the grass-root level and also create a revolutionary change in their daily, monotonous and conservative life style.

Empowerment of Women as a Unique Identity:

The concept of feminist movement under the canopy of women empowerment took its origin in the 18th Century in England. The women empowerment is not a magical process which can be achieved over the night. It can only be achieved with a continuous socio-economic progress. So, in the words of Doshi & Jain, it may be stated that "women are empowered through – women emancipation movement, education, communication, media, political parties and general awakening". Hence, the women empowerment is a continuous progress which requires constant efforts of Government, intellectual men and women who are concerned enough about enhancement of status of women in the society. Therefore, various socio-economic and political factors are there to facilitate the empowerment of women which can be discussed in the following manner:

i. Society must recognize the equal status of men and women;
ii. Women can become stronger only with educational and economic power because sufficient education can only help them to create an unique identity as well as it will make them less dependence on men;
iii. Women must have freedom to take care her own decision about her own life;
iv. They must be free to take part in the administrative process and political affairs;

Hence, in the process of active women empowerment we should start the process from the very grass-root level. For example it may be stated that, if we give sufficient opportunity to the women to take active participation in **Gram-Samiti** then they can able to raise their voice and can also establish their separate identity in the society. In this regard, the **Kultikri** village of **Paschim Midnapore (Sankhrail Block), West Bengal** deserves a special mention. That place has successively elected all women panchayats not because of the reservation for women. The mere fact is that its women leaders have beaten their male counter parts fair and square on the basis of their superior leadership qualities and remarkable track record of development.§§§§§§§§§§§§§§§§§ Therefore, the incident of this small village created a great revolution in the area of women empowerment as well as it will also encourage millions of women to create unique identity by promoting social transformation through a revolutionary approach.

Social Transformation of Women as a Separate Identity:

Women are the integral part of the society. Society can't be an ideal society without the contribution made by the women. In the matter of active women empowerment in the society the contribution of law deserves a special remark at national and international level. Therefore, various legal instruments which are responsible for the social upliftment of women as a separate identity can be described in the following manner:

---

§§§§§§§§§§§§§§§§§ Deccan Herald (Living She); Bangalore Edition; June 15, 2013.

Basic Human Rights Instruments for Women:

Since the Second World War, the advancement of rights women has been the most serious concerned of Women. So, the basic human rights which are available to women can be discussed in the following manner:

i. The Universal Declaration of Human Rights, 1948, provides that, all the rights and fundamental freedoms incorporated in the UDHR are available equally to both men and women without any distinction;

ii. The Declaration of Maxico on the Equality of Women and Their Contribution to Development and Peace, 1975, includes the following rights:
    ✓ Equality between men and women;
    ✓ Equal access to education and training;
    ✓ The right to work and equal pay for work of equal value;
    ✓ Right of every women to decide freely whether to marry (**etc**.)

iii. The Convention on the Political Rights of Women provides that:
    ✓ Women shall be entitled to vote in all elections on equal terms with men;
    ✓ Women shall be eligible for election to all public bodies, established by national law on equal terms with men without any discrimination.

Role of Constitution of India in the matter of Women Empowerment:

The various Constitutional Provisions are describing in the following manner:

- ❖ Art.14 provides that, "State shall not deny to any person equality before law and equal protection of laws within the territory of India". Hence, it may be stated that:

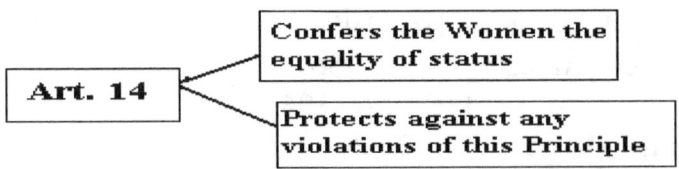

In *Madhu Kishwar & Others v. State of Bihar & Others******************** the Petitioners, members of HO and ORAN Tribes of Bihar had challenged the validity of Sec.7 and Sec.8 of the Chhota Nagpur Tenancy Act, 1908 and it was declared as the ultra vires of Art. 14 of the Constitution of India as Sec.7 and Sec.8 confines succession of property to the male and it was discriminating the women.

- ❖ Art. 15(3) provides that state shall make special provision for women and children;
- ❖ Art. 16 provides that, no discrimination be made by the State against its citizens including women while providing jobs;
- ❖ Right to Education under Art. 21-A††††††††††††††††††††;
- ❖ Art. 23 prohibits traffic in human beings and forced labour;
- ❖ Art.243-D (3) provides that, 1/3 members of seats shall be reserve for women in the Panchayats including the women belonging to SC's and ST's;

****************** (1992) 1 SCC 102

†††††††††††††††††††† Inserted by the Constitution(86th Amendment) Act, 2002

❖ Art.243-T provides that, 1/3 members of seats shall be reserve for women in the Municipalities including the women belonging to SC's and ST's. (etc.)

Empowerment of Women under Family Jurisprudence:

In the matter of empowerment of women under family jurisprudence the following legislations are playing the most cardinal role:

- The Hindu Marriage Act, 1955;
- The Hindu Succession Act, 1956;
- The Hindu Adoption & Maintenance Act, 1956;
- The Hindu Minority & Guardianship Act, 1956;
- The Criminal Procedure Code, 1973 (Section 125);
- The Prohibition of Child Marriage Act, 2006;
- The Muslim women Protection of Rights on Dowry Act 1986
- The Indian Divorce Act, 1969 (4 of 1969); (etc.)

1. Welfare of Women under through various Legal Instruments:

In order to ensure equal justice and to promote the momentum of women empowerment, the following Legislations deserve special mention:

- The Commission of Sati (Prevention) Act, 1987 (3 of 1988);
- Protection of Women from Domestic Violence Act, 2005;
- The Indecent Representation of Women (Prohibition) Act, 1986;
- The Sexual Harassment of Women at Workplace (PREVENTION, PROHIBITION and REDRESSAL) Act, 2013;
- The Immoral Traffic (Prevention) Act, 1956;
- The Dowry Prohibition Act, 1961 (28 of 1961) (Amended in 1986);
- The Indian Penal Code, 1860;

- The Workmen's Compensation Act, 1923;
- The Trade Unions Act 1926;
- The Payments of Wages Act, 1936;
- The Factories Act, 1948;
- The Maternity Benefit Act, 1961 (53 of 1961);
- The Medical Termination of Pregnancy Act, 1971 (34 of 1971);
- The Pre-Natal Diagnostic Techniques (Regulation and Prevention of misuse) Act 1994; (etc.)

Establishment of National Commission for Women:

The Central Government established the National Commission for women in the year of 1992 in order to check incidents violence against women as well as to promote social, legal and economic equality of women. The Commission consists of a Chairman, Five Members and a Member Secretary and all are nominated by the Central Govt. The Commission performs the following activities:

- Review of Legislations;
- Remedial Action to safeguard the interest of Women;
- To accord highest priority to secure speedy justice to women;
- To monitor the enforcement of Laws;
- To inquire into "Unfair Practice" against the women;

Compensation Jurisprudence under Art. 21 and Women:

Article 21 of the Constitution of India is the mother of all the Fundamental Rights. According to Article 21, "no person shall be deprived of his life and liberty except according to the procedure established by Law." Therefore, Article 21 can be divided into two parts i.e.

| Art.21 | = | Right to life + Personal Liberty |

This scope Art.21 is undoubted expanded after the historic decision of *Maneka Gandhi* ####################. Hence the compensation jurisprudence under Art.21 can be discussed with reference to judicial pronouncements in the following manner.

- In ***Delhi Domestic Working Women's Forum v. Union of India***,§§§§§§§§§§§§§§§§§§§§§ The Supreme Court had laid down the following guidelines.
  i. She must be given free legal aid;
  ii. Every Police Station must of list of Advocates;
  iii. She must be given free medical aid;
  iv. Name of the Victim shall not be disclosed;
  v. Proceedings should be in camera trail;
  vi. She must be rehabilitated;
  vii. She must be given interim compensation and there should be a Compensation Board;

- In ***Chairman, Railway Board v. Chandrima Das***********************, the High Court awarded a sum of Rs.10 Lakhs as compensation to the victim.

- In ***Bodhisathwa Goutam v. Subbra Chakraborty***††††††††††††††††††††††, the Supreme Court awarded an interim compensation of Rs.1000/- per month to the victim of Rape until her charges of rape are decided by the Trial Court.

- In ***Vishaka v. State of Rajastan***‡‡‡‡‡‡‡‡‡‡‡‡‡‡‡‡‡‡‡‡‡, the Supreme Court has laid down guidelines to prevent sexual harassment of working women in places of their work and the Court also held that, it is the duty of the employer or other responsible person in work place or other institutions, whether public or private to prevent sexual harassment of working women. The Court further stated that instances of sexual harassment resulting in violation of Fundamental Rights of women workers under Art.14, 19 & 21 of the Constitution of India.

---

‡‡‡‡‡‡‡‡‡‡‡‡‡‡‡‡‡‡‡‡‡ Maneka Gandhi v. Union of India, AIR 1978 SC 597
§§§§§§§§§§§§§§§§§§§ (1995) 1 SCC 14
********************* AIR 2000 SC 988
††††††††††††††††††††† (1996) 1 SCC 490
##################### AIR 1997 SC 3011

Hence in the words of Mr. Justice Saghir Ahmad it may be stated that "Women also have the right to be respected and Reactive as equal citizens. Their honor and dignity cannot be touched and violated. They also have the right to lead an honorable and peaceful life.

Legal Aid and Women Empowerment:

Legal aid is a key instrument to accelerate the process of illuminating every aspect of social justice in every corner of society. It is also considered as a dynamic instrument to achieve justice. The concept of legal aid also enhances the true spirit of "Rule of Law". Free legal aid is necessary because most of the women are there mainly in rural and semi urban area who cannot afford legal representation in the courts due to the financial difficulties and lack of proper knowledge. Therefore, in order to spread the concept of legal- aid in the every corner of society, the Constitution of India has taken special initiative. In Article 39-A which directs the State to ensure equal justice and free legal aid to economically backward class. This Article was mainly added to the Constitution pursuant to the new policy of the Government to give legal aid to the economically backward classes of people. Now, the Right to free legal aid and speedy trial are also guaranteed as fundamental right under Article 21. Hence, in order to achieve the objectives in art. 39-A, the State must encourage and support the participation of voluntary organizations or social action groups in operating a legal aid program.

Status of Dalit Women: [11]

The vulnerability of *Dalit* women as depicted by a Nepali writer can be taken as an example on how these women are oppressed in the total social and family context. In her story of A *Naikape Sarkilli,* Parijat has well described the story of a low caste woman who has to earn her living by digging sand in the Bagmati River, in Kathmandu valley. The story goes like this —

The woman works so hard digging the sand in Bagmati River during the cold and chill winter of Kathmandu. During that cold winter she has wrapped herself in a thin saree and a blouse without proper winter clothes. She returns home after she finishes the work at 4 O'clock in the evening which is *almost* dark during the winter time. She then begins to cook the evening meal and waits for her drunken husband to come home from playing cards, which is his daily job. Her husband who is disabled cannot help her in anything even giving the physical comfort, but assaults her with bitter words and gets satisfied as being a man. He comes to the *pati* (shelter for the homeless) with all his frustration, inferiority *complex* and defeat which automatically comes in his words. He still demands the money she earned by her hard work as his right, to drink alcohol which she cannot protect without a proper place to hide. After completing all the household chores she sleeps in her wet clothes that is all she has to wear. "She is as cold as the sand of Bagmati, being deprived by everything". She is already turned to her old age in her middle age. She can neither revolt nor fight against this injustice.

This is a typical story described by the writer which represents the condition of majority of the *Dalit* women's life. They have to live a miserable life be it social, economic educational, legal and others.

11. *http://himalaya.socanth.cam.ac.uk/collections/journals/opsa/pdf/OPSA_08_06.pdf browsed on 14.12.2013* at about 8 p.m.

Women in Political Participation:

Women have the equal right to participate in the process of Parliamentary democracy actively beside a man because in every modern political system election process occupies the most cardinal role. In many studies and research it is found that women, particularly of backward caste and class in the rural areas are not independent voters due to the following reasons:

- That majority of them are illiterate;
- That majority of them don't have enough independence to express their views and they make their choice on the basis of suggestions given by male members of the family, viz. husband, and sons;
- They lack information and political awareness;
- They are not politically conscious.

Therefore, special initiative must be taken to increase the number of women voters, particularly of backward caste through proper education and awareness program. It is necessary to send the message to those women that their life is not confined within the four walls of kitchen and other house-hold works, they must come forward to participate not only in democratic process but also in administrative process, legal process, educational process etc. to enhance their entity in the society. Hence, it may be stated that:

- ✓ Article1 of the Convention on the Political Rights of Women, 1952 provides that, the women shall be entitled to vote in all elections on equal terms with men;

- ✓ Article 2 of the Convention on the Political Rights of Women, 1952 provides that, women shall be eligible for election to all publicly elected bodies, established by national law, on equal terms with men and without any discrimination;
- ✓ In India, Women's Reservation Bill or the Constitutional (108th Amendment) Bill, is a pending Bill which proposes to amend the Constitution of India to reserve 33% of all seats in the Lok-Sabha and in all State Legislative Assemblies for women;
- ✓ Art. 243-D and Art. 243-T provides that in every Panchayats and Municipalities 1/3 members of seats shall be reserve for women including the women belonging to SC's and ST's. (etc.)

Conclusion and Suggestive Measures:

Thus, from the above discussion it may be concluded that that law has been proved as an important instrument for the development of the status of women and it also brings a new hope for millions of oppressed women in India. So, the following suggestive measures may be considered for the betterment of social status of women in the society under the canopy of socio-legal aspect:

- ➢ More attention must be given on women education by establishing educational institutions in rural, semi urban and urban areas;
- ➢ Application of legal instruments must be strengthening in order to eliminate crimes against the women;
- ➢ It is necessary to conduct the awareness program to eliminate gender biasness at every *Gram-panchayat* level in rural area;
- ➢ Special attention must be given to improve conditions of health centres in rural as well as urban areas specially for women and these health centres must spread the good advices for the betterment of healthy life to women by conducting various workshops and health awareness camps at least twice a month;
- ➢ It is necessary to establish legal aid camp at every sub-divisional and *Gram-panchayat* level to provide necessary legal assistant to women.
- ➢ Women panchayat must be established to encourage the leadership qualities;
- ➢ NGOs must actively take special efforts to encourage and to motivate the women especially in rural area to participate in active democracy and various administrative processes.

Hence, the advancement of women involves concerned commitment from men as well as women. It also indicates that, men and women both should jointly move to promote socio-economic development of those women who are socially backward and economically week. So, welfare is necessary because a nation's development can be measured by looking at the social condition of women in that country. Therefore, it can also be stated that, men and women are like two flowers in a same twig. The social transformation beckons the modernization of law when the society needs freedom from its orthodox ideology which hinders the decent movement of the society towards a welfare state.

References

1. Deccan Herald (Living She); Bangalore Edition; June 15, 2013
2. Dr. Pandey, J.N., The Constitutional Law of India; Central Law Agency, Allahabad; Edition: 2005 & 2012.
3. Dr. Chandra, U; Human Rights; Allahabad Law Agency Publication, Allahabad; Edition: 2005.
4. Agarwala, R.K; Hindu Law (Codified & Uncodified); Central Law Agency, Allahabad, Edition: 2005.
5. Dr. P. Ishwar Bhat; Law & Social Transformation; Eastern Law Company, Lucknow; Edition:2009;
6. Agarwal, Nomita; Women and Law in India; New Century Publications, Delhi; Edition: 2002.
7. Afsar Bano; Women and Social Change; Reference Press, New Delhi; Edition: 2003.
8. Rao, Shankar C.N.; Sociology, Principle of Sociology with an Introduction to Sociological Thought; S. Chand & Company Ltd.; New Delhi; Edition: 2012.
9. Saikia, Sailajananda (2014). Caste Based Discrimination in the enjoyment of Fundamental Rights: A Critical Review of the present status of Dalit's in India. International Journal of Research (IJR). Vol-1, Issue-2.
10. http://ncw.nic.in/frmLLawsRelatedtoWomen.aspx)
11. http://en.wikipedia.org/wiki/Women%27s_Reservation_Bill

Synthesis and Characterization of Ni Doped Zno Nanoparticles

M.R.A. Bhuiyan & M.K. Rahman[a]

Abstract

This paper discerns key ideas and themes of the possibility of growing Ni doped ZnO nanoparticles by electrochemical method. The purpose is to study the growth mechanism and to optimize the parameters of this method. Upon successful synthesizing the samples, they were characterized using various techniques. XRD, SEM, FTIR, photoluminescence spectroscopy together with the measured optical parameters obtained from UV-VIS absorption testing were analyzed. The X-ray diffraction (XRD) was measured by using a Bruker D8 Advance X-ray diffractometer with CuK_α radiation. The surface morphology was investigated using an 'EVO LS 15' scanning electron microscope. The FTIR absorption spectra were recorded on a Perkin-Elmer GX FTIR system. The PL spectra were collected on a Jobin Yvon-Horiba Triax 190 spectrometer with a spectral resolution of 0.3 nm. UV-VIS absorption spectrum was recorded by using a UV-VIS spectrophotometer in the photon wavelength range between 300 and 600 nm. XRD pattern reveals that the polycrystalline of hexagonal wurtzite structure and the average size of the particles were estimated to be approximately 61 nm, which conform the nanoparticle. The FTIR result shows the stretching vibration of the Zn-O bond in Ni doped ZnO nanoparticles. There is a green emission peak centered at about 384 nm in the PL behavior. The band edge is shifted to the lower energy side of the Ni doped ZnO nanoparticle. Analyzing the results of various types of characterizations, it has been assessed that Ni doped ZnO nanoparticles was successfully synthesized.

Keywords:

Ni doped ZnO, Compositional uniformity, Structural parameter, Optical parameter.

---

[a] *Department of Applied Physics, Electronics and Communication Engineering, Islamic University, Kushtia-7003, Bangladesh*
E-mail: mrab_iu@yahoo.com [a*] *Author for Correspondence*: E-mail: mkr_iu@yahoo.com

For Referring this Paper

Bhuiyan, M.R.A & Rahman, M.K. (2014). Synthesis and Characterization of Ni Doped Zno Nanoparticles. *International Journal of Research (IJR)*. Vol-1, Issue-3. Page 67-73. ISSN 2348-6848.

INTRODUCTION

Nanometer-sized particles (1–100 nm) have attracted considerable interest for a wide variety of applications, ranging from electronics via ceramics to catalysts due to their unique or improved properties, which are primarily determined by size, composition, and structure [1]. These properties are strongly related to the synthesis processes. Numerous solution techniques, for example, sol–gel, emulsion, colloidal, and aerosol processes, have been used to synthesize a variety of nanoparticles [2–6]. As an alternative approach, the electrochemical route is of considerable interest because of a possibly precise particle size control achieved by adjusting current density or applied potential. For electrochemical synthesis, extensive investigations have been focused on the metal particles, especially on noble metal particles. This method, in fact, skillfully combines an electrochemical process with plasma at ambient pressure and temperature. Compared with other methods of synthesizing Ni doped ZnO powder, this access has some advantages. Firstly, the whole process, with a simple experimental set-up and electrolyte system is performed under mild conditions. Secondly, particles can be obtained within a few minutes. Finally, changing the volume ratio of the electrolyte can effectively control the size of the obtained particles. The earlier study of the magnetic nature of Ni doped ZnO shows different results. Some authors [7, 8] reported that, ZnO doped with 5 to 25at. % Ni did not show room temperature ferromagnetic (RT FM) behaviour and the FM characteristics disappeared for temperatures higher than 30K. Wakano et al. [9] observed and measured FM properties in 3 to 25 at.% Ni doped ZnO at temperatures less than 30K, however, super paramagnetic behaviour was observed at least upto 300K. i.e. absent of RT-FM. Liu et al. [10] have noticed that the preparation details of the prepared Ni doped ZnO has great influence on their magnetic properties, so that annealing at 800°C in Ar gas atmosphere significantly increases the magnetization, however, RT-FM was not detected. Recently, the RT-FM was observed in ~12at. % Ni doped ZnO prepared by sputtering method [11] and in 5at. % Ni doped ZnO prepared by laser ablation method on sapphire substrate [12]. Thus, the experimental evidence for the existence of RT-FM phase in Ni doped ZnO at room temperature is still questionable and needs more investigations. Generally, experiments show that the magnetization properties of ZnO:Ni depend strongly on the preparation method and procedure and seems to be sensitive to the crystalline structure and concentration of intrinsic (like oxygen vacancies) and extrinsic defects. A clear understanding of the RT-FM ordering can provide us with new experimental approaches to an opportunity to develop spintronic devices based on Ni doped ZnO.

EXPERIMENTAL DETAILS

Synthesis of Ni doped ZnO Nanoparticles

The synthesis method has been employed a modified version of the originally used by Reetz and Helbig [13] for metal particles (in a one-phase electrochemical system). Ni doped ZnO particles were synthesized at room temperature by an electrochemical route. The electrolytic bath consisted of acetonitrile and tetrahydrofuran (THF) mixed in the ratio 4:1, in which high purity Ni doped Zn metal sheet (1cm × 1cm) and laboratory grade platinum sheet (1cm × 1cm) served as anode and cathode respectively. The capping agent tetra-trimethyle ammonium-bromide (TTAB) also served as the electrolyte. Electrolysis was carried out in nitrogen atmosphere for a few hours in constant current mode (GPS-30D, 0-30V, 0-5A). Current density was maintained to obtain different sized of particles. The molarity of TTAB in the chemical bath was varied from 0.1 mM to 0.9 mM. The white Ni doped ZnO particles remain suspended in the solvent and separated by using centrifugation. On drying, a free flowing powder of Ni doped ZnO particles are obtained.

Measurements

The structural properties of the Ni doped ZnO particles were measured by using a Bruker D8 Advance X-ray diffractometer with CuK$_\alpha$ radiation of wavelength λ = 1.54056 Å. The X-ray diffraction (XRD) measurements were carried out in the locked coupled mode in the 2θ range of 20 to 60°. The surface morphology and composition of Ni doped ZnO particles were investigated by using an 'EVO LS 15' scanning electron microscope developed by Carl Zeiss. An accelerating voltage of 15 to 19 keV and probe current of ~800 pA. UV-VIS absorption spectrum of Ni doped ZnO particles were recorded by using a UV-VIS spectrophotometer in the photon wavelength range between 300 and 600 nm. The FTIR absorption spectra were recorded on a Perkin-Elmer GX FTIR system used to obtain 16 cm^{-1} resolution spectra in the range 500 to 4000 cm^{-1} region, scanned 30 times (absorbance mode), in order to exploit the instrumental built-up noise reduction algorithm. The Photoluminescence (PL) measurements were carried out from room temperature by employing a 488 nm line of an argon ion laser. The PL spectra were collected on a Jobin Yvon-Horiba Triax 190 spectrometer with a spectral resolution of 0.3 nm, coupled with a liquid nitrogen-cooled CCD detector.

RESULTS AND DISCUSSION

Structural and Compositional

X-ray diffraction (XRD) is mainly used for phase identification. Fig.1 shows the XRD pattern of the prepared Ni-doped ZnO sample. The grown sample shows the peaks of (100) and (101). No signals of the metallic Zn are detected by XRD. Also, there is no peak corresponding with the Ni, suggesting that the Ni element may be doped into ZnO. This pattern reveals that the polycrystalline of hexagonal wurtzite structure that is known ZnO structure. The

Bragg angle of the intense (101) reflection is observed as light shift towards higher values relative to that of pure ZnO which has been indicated that Ni-doped ZnO was formed along with NiO phase. Of course, this happens relating to the limit of the solids solubility of Ni in ZnO. This is an evidence for creation of internal compressive micro stress. Such case was also observed in ZnO annealed in hydrogen atom sphere [14]. It is known that more oxygen content is introduced into the sample. It is clear that the (101) peak is sharper and stronger.

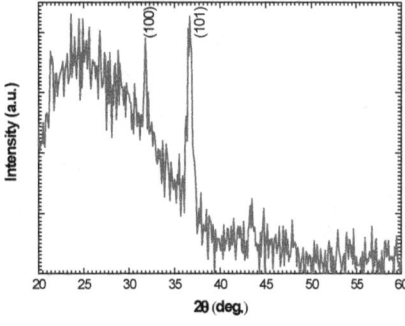

Fig. 1. XRD spectrum of Ni doped ZnO particles

The size of the Ni doped ZnO particles was estimated by applying the Scherrer equation [15] to the half intensity width of the (101) peak:

$$d = \frac{k\lambda}{\beta \cos\theta}$$

(1)

where k is the particle shape factor and taken as 0.827 because of the hexagonal Ni doped ZnO particles, λ is the wavelength of CuKα radiation (0.154nm), β is the calibrated half intensity width of the selected diffraction peak (degrees), and θ is the Bragg angle (half of the peak position angle). From this equation, the average size of the Ni doped ZnO particles was estimated to be approximately 61 nm, which conforms reasonably well to the literature value [16]. This implies that the samples are successfully synthesized.

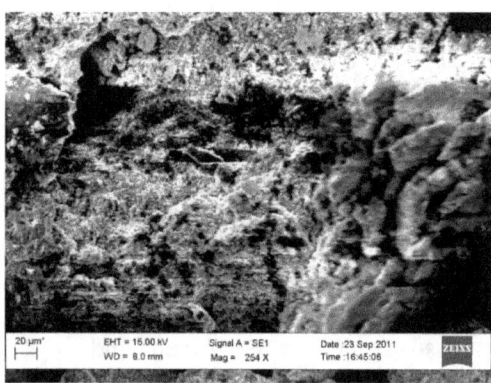

Fig. 2. SEM image of Ni doped ZnO particles

SEM image in Fig. 2 shows the morphology of Ni doped ZnO particles. It is observed that there is a rough surface in the particles.

Optical absorption

Substation of Ni cations in tetrahedral sites of the wurtzite structure was further conformed by UV–vis optical spectroscopy. The room temperature spectra of the Ni doped ZnO particles are reported in Fig. 3 and compared to the spectrum recorded for the other researcher report [17].

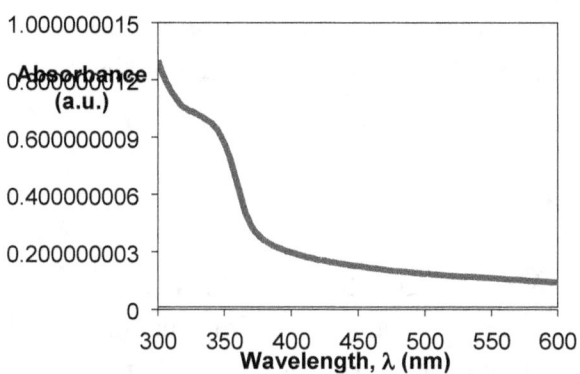

Fig. 3. Optical absorption spectrum of Ni doped ZnO particles

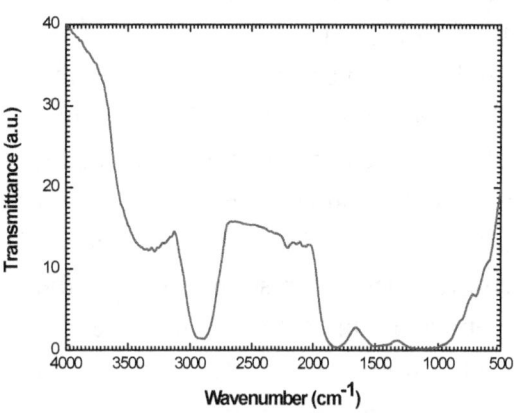

Fig. 4. FTIR spectrum of Ni doped ZnO particles

From this figure, the band edge is shifted to the lower energy side of the Ni doped ZnO samples. The decrease in the band edge is a clear indication for the incorporation of Ni inside the ZnO lattice [18].

FTIR absorption

To make the clear interactions with the particles, FTIR absorption spectrum of the particles was performed. Fig. 4 shows the FTIR spectrum of Ni doped ZnO particles. The peak at 3180 cm^{-1} is the stretching vibration of the H-O bond. The peaks at 1620 and 1400 cm^{-1} are assigned to the vibrations of amide I and amide II, respectively. The peaks are similar to other worker [19]. A peak at 530 cm^{-1} is the stretching vibration of the Zn-O bond in doped ZnO particles.

Photoluminescence spectroscopy

The room-temperature PL spectrum on the Ni doped ZnO sample is shown in fig. 5. Only a UV emission

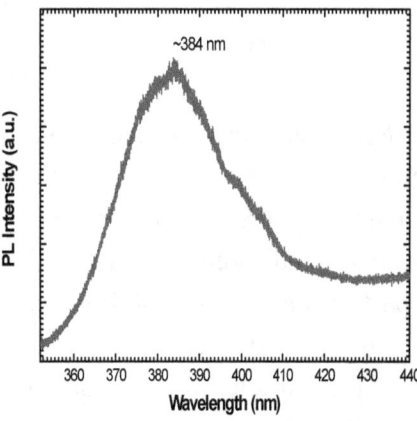

Fig. 5. Photoluminescence spectrum of Ni doped ZnO particles

peak can be observed on the pure ZnO cone arrays without any significant emission in visible region, though the green emission is often observed in many reports [20-23]. The UV emission is attributed to the near-band-edge emission from the recombination of free excitons through an exciton-exciton collision process [24]. Vanheusden et al. [22] proved that the individually ionized oxygen vacancy is responsible for the green emission in ZnO. To observe the green emission that indicates its good crystallization with little oxygen vacancy. After the Ni ion incorporation, the UV emission peak shifts to a little longer wavelength, consistent with the result of the UV-visible absorption. Besides, the intensity of UV peaks of Ni doped ZnO decreases. There is a green emission peak centered at about 384 nm. As more oxygen ions would be pulled out from lattices by Ni ions to form NiO in interstices of lattices.

CONCLUSIONS

In summary, Ni doped ZnO particles have been synthesized by a simple electrochemical route at room temperature. Structure analysis indicated that the Ni doped ZnO particles are polycrystalline of hexagonal wurtzite structure and the average size of the particles was estimated to be approximately 61 nm, which conforms the nanoparticles in form. The band edge is shifted to the lower energy side of the nanoparticles. The FTIR result shows that the stretching vibration of the H-O and Zn-O bond. Photoluminescence spectrum measurement demonstrated that the nanoparticles exhibit a strong near band edge UV emission peak centred at 384 nm.

ACKNOWLEDGEMENTS

We are grateful to Ministry of Science and Information, Communication Technology authority for financial support. We are thankful to Dr. M. Monjarul Alam, Mr. M. Abdul Momin for their assistance in preparing and measurement the samples.

REFERENCES

[1] Xia B, Lenggoro W, Okuyama K. Novel route to nanoparticle synthesis by salt-assisted aerosol decomposition. Adv Mater 2001; 13:1579-82.

[2] Lakshmi BB, Dorhout KP, Martin RC. Sol-gel template synthesis of semiconductor nanostructures. Chem Mater 1997; 9:857-62.

[3] Kobayashi S, Hanabusa K, Suzuki M, Kimura M, Shirai H. Preparation of TiO_2 fiber in a sol-gel system containing organogelator. Chem Lett 1999; 10:1077-78.

[4] Kobayashi S, Hanabusa K, Hamasaki N, Kimura M, Shirai H, Shinkai S. Preparation of TiO_2 hollow-fibers using supramolecular assemblies. Chem Mater 2000; 12: 1523-25.

[5] Rao RNC, Satishkumar CB, Govindaraj A. Ziconia nanotubes.Chem Commun 1997; 16:1581-82

[6] Satish Kumar CB, Govindaraj A, Nath M, Rao NC. Synthesis of metal oxide nanorods using carbonnanotubes as templates.J Mater Chem 200; 10:2115-19.

[7] Li T, Qiu H, Wu P, Wang M, Ma R. Characteristics of Ni-doped ZnO: Al films grown on glass by direct current magnetron co-sputtering. Thin Solid Films 2007; 515:3905-09.

[8]   Ueda K, Tabata H, Kawai T. Magnetic and electric properties of transition-metal-doped ZnO films.

Appl Phys Lett 2001; 79:988-90.

[9]   Wakano T, Fujimura N, Morinaga Y, Abe N, Ashida A, Ito T. Magnetic and magneto-transport

Properties of ZnO:Ni films. Physica E 2001; 10:260-64.

[10]   Liu E, Xiao P, Chen JS, Lim BC, Li L. Ni doped ZnO thin films for diluted magnetic semiconductor materials. Current Appl Phys 2008; 8:408-11.

[11] Pandey B, Ghosh S, Srivastava P, Avasthi DK, Kabiraj D, Pivin JC. Synthesis and characterization of Ni doped ZnO: A transparent magnetic semiconductor. J Mag Mat 2008; 320:3347-51.

[12] Thota S, Kukreja LM, Kumar J. Ferromagnetic ordering in pulsed laser deposited $Zn_{1-x}Ni_xO$/ZnO bilayer thin films. Thin Solid Films 2008; 517:750-54.

[13] Reetz MT, Helbig W. Size-selective synthesis of nanostructured transition metal clusters. J Am Chem Soc1994; 116:7401-02

[14] Oh BY, Jeong MC, Kim DS, Lee W, Myoung JM. Post-annealing of Al-doped ZnO films in hydrogen atmosphere. J Cryst Growth 2005; 281:475-80.

[15]   Mahmoud WE, Al-Ghamdi AA, El-Tantawy F, Al-Heniti S. Synthesis, characterization and charge transport mechanism of CdZnO nanorods. J Alloys Compd 2009; 485:59-63

[16]   Al-Harbi T. Hydrothermal synthesis and optical properties of Ni doped ZnO hexagonal nanodiscs. J Alloys Compd 2011; 509:387-90.

[17] Cong CJ, Hong JH, Liu QY, Liao L, Zhang KL. Synthesis, structure and ferromagnetic properties of Ni-doped ZnO nanoparticles. Solid State Communications 2005; 138:511-15.

[18] Radovanovic PV, Gamelin DR. High-temperature ferromagnetism in Ni^{2+} doped ZnO aggregates prepared from colloidal diluted magnetic semiconductor quantum dots. Phys Rev Lett 2003; 91:157202-205.

[19] Wang Y, Liao X, Huang Z, Yin G, Gu J, Yao Y. Preparation and characterization of Ni-doped ZnO particles via a bioassisted process. Coll and Surf A: Physchem Engg Aspects 2010;372:165-71.

[20] Kong YC, Yu DP, Zhang B, Fang W, Feng SQ. Ultraviolet-emitting ZnO nanowires synthesized by a physical vapor deposition approach. Appl Phys Lett 2001;78:407-09.

[21] Yang JH, Wang DD, Yang LL, Zhang YJ, Xing GZ, Lang JH, Fan HG, Gao M, Wang Y. Effects of supply time of Ar gas current on structural properties of Au-catalyzed ZnO nanowires on silicon (1 0 0) grown by vapor–liquid–solid process. J Alloys Compd 2008;450:508-11.

[22] Vanheusden K, Seager CH, Warren WL, Tallant DR, Voigt JA. Correlation between photoluminescence and oxygen vacancies in ZnO phosphors. Appl Phys Lett 1996;68:403-05.

[23] Vanheusden K, Warren WL, Seager CH, Tallant DR, Voigt JA, Gnade BE. Mechanisms behind green photoluminescence in ZnO phosphor powders. J Appl Phys 1996;79:7983-90.

[24] Schwartz DA, Kittilatved KR, Gamelin DR. Above-room-temperature ferromagnetic Ni^{2+} doped ZnO thin films prepared from colloidal diluted magnetic semiconductor quantum dots. Appl Phys Lett 2004;85:1395-97.

The Place of Customary International Law in the Nigerian Legal System – A Jurisprudential Perspective

C.J.S. AZORO, Esq.,[85]

Abstract

Every society has a framework of laws and principles upon which it develops. The international society thus posits various rules upon which the sovereign states and other subjects of international law may develop in pursuit of the actualization of their interests. A similar situation obtains in Nigeria where her legal system prescribes various laws towards regulating social relations within her jurisdiction. As a sovereign state, Nigeria remains subject to international law with the incidental international responsibility for any breach of same. Though her legal system allows for the enforcement of international treaties in her municipal courts subject to certain qualifications, the law appears to be silent on the status of customary international law. This paper argues that customary international law forms part of the Nigerian legal system and should be applied where appropriate towards the maintenance of peaceful co-existence between all interests represented in the Nigerian society.

Keywords:

Individual, Law, Jurisdiction, Custom, State.

For Referring this Paper:

C.J.S. AZORO, Esq (2014). The Place of Customary International Law in the Nigerian Legal System– A Jurisprudential Perspective. *International Journal of Research (IJR)*. Vol-1. Issue-3. Page 74-100. ISSN 2348-6848.

---

[85] LL.B (NAU), B.L., B.C. (WONMMC), Barrister and Solicitor of the Supreme Court of Nigeria, +234 806 111 693 5; +234 811 61 234 36, cjsazoro@yahoo.com

Introduction

The move from individuality to communality in the history of human evolution cum civilization lends credence to the proposition that friction and tension are necessarily incidental to social interaction and existence. Indeed, it is undoubtedly true that human existence is founded on and sustained by the conceptualization of law, as a natural, physical and social phenomenon. This is based on the fact that law, as an instrument of social engineering, plays the primary role of striking a balance between the multifarious competing interests represented in the society.

Thus, on the municipal level, law maintains an important balance between the interests and rights of the individuals *inter se*, and that of the individual *vis-à-vis* the state. Indeed, the position in Nigeria is aptly stated by section 1(1) of the Constitution,[86] to the effect that its provisions is the supreme law and is binding on all persons and authorities (including Nigeria itself as a State).[87]

Similarly, in the international plane, law also maintains the necessary balance between the interests cum rights of the various subjects of international law. Indeed, the purposes of international law include: resolution of problems of a regional or global scope, regulation of areas outside the control of any one nation and adoption of common rules for multinational activities. International law also aims to maintain peaceful international relations and resolve international tensions peacefully when they develop, to prevent needless suffering during wars, and to improve the human condition during peacetime.[88]

However, the concept of sovereignty entitles the State to determine what laws should obtain within her jurisdiction,

---

[86] 1999 Constitution of the Federal Republic of Nigeria (as amended), hereinafter referred to as the 1999 CFRN (as amended).

[87] *African Continental Bank Plc* v *Losada (Nig.) Ltd* [1995]7 NWLR (pt 405) 26; *Lakanmi* v *Attorney-General of Western State* (1971)1 UILR 210, (1974) ECSLR 713; *Ereku* v *Military Governor, Midwestern State* (1974)10 SC 59; *Onyiuke* v *Eastern States Interim Assets and Liabilities Agency* (1974)10 SC 77. This position is given more impetus by section 1(3) of the 1999 CFRN (as amended) which provides that "if any law is inconsistent with the provisions of the Constitution, the Constitution shall prevail and that other law shall to the extent of the inconsistency be void" - *Kalu* v *Odili* [1992]5 NWLR (pt 240) 130; *Phoenix Motors Ltd* v *NPFMB* [1993]1 NWLR (pt 272) 718; *Speaker, Kogi State House of Assembly & 4 Ors* v *Hon. David Adegbe* [2010]10 NWLR (pt 1201) 45; *Cadbury Nig. Plc* v *Federal Board of Inland Revenue* [2010]2 NWLR (pt 1179) 561; *A.G - Ondo State* v *A.G - Federation* [2002]9 NWLR (pt 702) 222; *Hope Democratic Party* v *Mr. Peter Obi & 5 Ors* [2011]18 NWLR (pt 1278) 80; *Ekulo Farms Ltd* v *Union bank of Nigeria Plc* [2006] All FWLR (pt 319) 895; *Fasakin Foods (Nig.) Ltd* v *Shosanya* [2006] All FWLR (pt 320) 1059.

[88] E.A. Oji, "Application of Customary International Law in Nigerian Courts", *NIALS Law and Development Journal* (2011) 1(1), p 151.

without prejudice to her international responsibility with reference to breaches of international law. Nigeria is no exception such that section 12 of her Constitution provides that *"no treaty between the Federation and any other country shall have the force of law except to the extent to which any such treaty has been enacted into law by the National Assembly"*.

The implication of the above is that international law is recognized by the Nigerian legal system provided it is codified in a treaty to which Nigeria is a party and has been domesticated by an Act of the National Assembly.[89] No mention is made of customary international law and it appears that it is inapplicable in so far as Nigerian jurisprudence is concerned. This paper makes a critical analysis of the concept of customary international law, its relationship with municipal law and the Nigerian legal system. It pontificates that customary international law is recognized by the present state of the Nigerian legal system and makes a case for its application in Nigeria. It enjoins the judiciary (both the Bar and the Bench) to adopt same where applicable in the determination and resolution of disputes; just as it calls upon the legislature and the executive to take cognizance of same in the making and implementation of laws and policies for the sustainable development of the Nigerian society.

Sources of International Law

Any discourse on the sources of international law should rightly start from the provisions of article 38(1) of the 1945 Statute of the International Court of Justice (ICJ) which provides that:

> The Court, whose function is to decide in accordance with international law such disputes as are submitted to it, shall apply: (a) international conventions, whether general or particular, establishing rules expressly recognized by the contesting states; (b) international customs, as evidence of a general practice

---

[89] *General Sani Abacha* v *Gani Fawahinmi* [2000] FWLR (pt 4) 533 at 585-586, [1996]9 NWLR (pt 475) 710; *Comptroller of Nigerian Prisons v Adekanye* [1999]10 NWLR (pt 623) 400; *Ubani v. Director, State Security Service* [1999] 11 NWLR (pt 625) 129; Also see the High Court decisions in *Mohammed Garuba & Ors v Attorney-General of Lagos State & Ors* (unreported judgment of the Lagos State High Court in Suit No. LD/559M/90), *Bamidele Opeyemi & Ors v Professor Grace Alele-Williams* (unreported judgment of the Bendel State High Court in Suit No. B/6M/89) and *Gani Fawehinmi v The President* (Unreported, Suit No. M/349/92), all cited in J.A. Dada, "Human Rights under the Nigerian Constitution: Issues and Problems", *International Journal of Humanities and Social Science* [Special Issue - June 2012]2 (12), pp. 33 – 43.

accepted as law; (c) the general principles of law recognized by civilized nations; (d) …judicial decisions and the teachings of the most highly qualified publicists of the various nations, as subsidiary means for the determination of rules of law.

This provision is widely recognized as the most authoritative and complete statement as to the sources of international law.[90]

International conventions, otherwise called treaties, are written agreements between two or more sovereign states. International organizations may also be given the capacity to make treaties, either with sovereign states or other international organizations.[91] Although treaties are basically agreements between the parties thereto, the binding effect conferred on them by the various enforcement machineries in the international sphere imbues same with a status akin to legislation at municipal law. Treaties may incorporate rules of custom or develop new law. Treaty law thus is created by the express will of states. The present system of international law remains largely consensual and centered on the sovereign state. It is within the discretion of each state to participate in the negotiation of, or to sign or ratify, any international treaty. Likewise, each member state of an international organization such as the United Nations is free to ratify any Convention adopted by that organization.[92]

Customary international law on its part is unwritten and derives from the actual practices of nations over time. To be accepted as law, the custom must be long-standing, widespread and practiced in a uniform and consistent way among nations. Some customary international law has been codified in recent years.[93] For example, the Vienna Convention on the Law of Treaties codified the customary law principle of *pacta sunt servanda* which is to the effect that treaties between sovereign states are binding on the parties thereto and must be followed in good faith.[94] However, it

---

[90] I. Brownlie, *Principles of Public International Law*, 6th edn, Oxford, Oxford University Press, 2003, p. 5; M.O. Hudson, *The Permanent Court of International Justice: A Treatise*, New York, Macmillan, 1934, p. 601. According to Professor Shaw, *"In international law, it is a dynamic source of law in the light of the nature of the international system and its lack of centralized organs"* – M.N. Shaw, *International Law*, 6th edn, New Delhi, Cambridge University Press, 2008, p. 70.

[91] E.A. Oji, *op cit*, p. 152.

[92] *Ibid*. It has been said that "the treaty making process is a rational and orderly one, permitting participation in the creation of law by all states on the basis of equality" – Henkin *et al*, *International Law: Cases and Materials*, St. Paul-Minnesota, West Publishing Co., 1980, p. 73.

[93] E.A. Oji, *op cit*, p. 154.

must be stated that treaties are not superior to customary law so that the former does not necessarily override the later and may co-exist with it.[95]

The phrase 'general principles of law recognized by civilized nations' is taken to connote principles so general as to apply within all systems of law that have achieved a comparable state of development.[96] According to Professor Shaw, situations do arise where there is no treaty, custom or judicial authority to cover a particular point in international law. It is for such a reason that the general principles of law recognized by civilized nations came to be recognized as a source of international law so as to close such gap which is otherwise known as *non liquet*.[97] Some of the general principles of law that has been applied by the courts include:

i. the principle of *res judicata*;[98]
ii. the principle that any breach of obligation incurs liability to make reparation;[99]
iii. the general principle of subrogation;[100]
iv. the principle of lifting the veil;[101]
v. the principle of circumstantial evidence;[102]
vi. the general doctrines of equity.[103]

Judicial decisions rendered by international courts are important elements in identifying and confirming international legal rules.[104] The most important international courts are the International Court of Justice, which mainly handles legal disputes between nations, and the International Criminal Court, which prosecutes individuals for genocide, war crimes and other serious crimes of international concern.[105] It must however be stated that although several regional courts have been established,[106]

---

[94] 1969 Vienna Convention on the Law of Treaties, article 26; *Gabcikovo-Nagymaros Project Case (Hungary v Slovakia)*, ICJ Reports, 1997, p. 7; *Legality of the Threat or Use of Nuclear Weapons Case (Advisory Opinion)*, ICJ Reports, 1996, p. 102; *Nicaragua v USA (Military and Paramilitary Activities in and against Nicaragua)* Case, ICJ Reports, 1986, p. 392.
[95] *Ibid*.
[96] Henkin *et al, op cit*, p. 75.
[97] M.N. Shaw, *op cit*, p. 98; H. Thirlway, "The Law and Procedure of the International Court of Justice", *BYIL* (1988), p. 76; P. Weil, "The Court Cannot Conclude Definitely...? *Non Liquet* Revisited", *Columbia Journal of Transitional Law* (1997)36, p. 109; E.A. Oji, *op cit*, p. 154 – 155.
[98] *Genocide Convention (Bosnia and Herzegovina v Serbia and Montenegro)* Case ICJ Reports, 2007, p. 113.
[99] The *Chorzow Factory* Case (1928) PCIJ Series A No. 17 p. 29.
[100] *Mavrommatis Palestine Concessions* Case (1924) PCIJ Series A No. 2 p. 28.
[101] The *Barcelona Traction* Case ICJ Reports, 1970, p. 6 at 39.
[102] *Corfu Channel* Case *(U.K v Albania)* ICJ Reports, 1949, p. 4.
[103] *Diversion of Water from River Meuse (Netherlands v Belgium)* Case (1937) PCIJ Series A/B No. 70 p. 73.
[104] 1945 Statute of the International Court of Justice, article 38(1)(d).
[105] E.A. Oji, *op cit*.
[106] Examples include the European Court of Human Rights (ECHR), Inter-American Court of Human Rights; African Court of Human and

the principle of *stare decisis* does not apply to international judicial tribunals though their decisions are often cited and utilized in subsequent decisions.[107]

The opinions of highly qualified publicists as reflected in their writings also constitute a veritable source of international law. Starting from Gentili and Hugo Grotius, writers have contributed immensely to the development of international law.[108] Indeed, the works of such renowned writers constitute a means of ascertaining rules of customary international law. Such writings remain a way of arranging and putting into focus the structure and form of international law and of elucidating the nature, history and practice of the rules of international law. They play a useful role in stimulating thought about the values and aims of international law, as well as pointing out the defects that exist within the system, while making suggestions for the future.[109]

Resolutions and decisions of the United Nations and other international organizations now also have a great impact on the views and practices of sovereign states, sometimes leading to rapid formation of customary international law.[110] The activities of some of these organizations result in draft conventions that may be adopted as treaty by the United Nations General Assembly.[111] Indeed, the United Nations Security Council, and some international organizations such as the European Union, have been conferred with power to enact directly binding measures.[112]

The Nature of Customary International Law

Albeit at risk of prolixity, it must be stated that article 38(1) of the ICJ Statute[113] establishes the place of international customs as a source of international law.[114] The existence of customary rules of international law can be deduced from the practice and behavior of states.[115]

Peoples' Rights (ACHPR), etc.

[107] The position is aptly stated in article 59 of the 1945 Statute of the International Court of Justice to the effect that decisions of the court have no binding force except as between the parties and in respect of the case under consideration. However, the common practice is for the courts to follow earlier decisions unless the circumstances of the particular case under consideration suggest the contrary – *Cameroun* v *Nigeria* (Case Concerning the Land and Maritime Boundary between Cameroon and Nigeria) (Preliminary Objections) ICJ Reports, 1998, p. 275.

[108] M.N. Shaw, *op cit*, p. 112.

[109] *Ibid*, p. 113.

[110] E.A. Oji, *op cit*, pp. 155 – 156.

[111] For example, the International Law Commission – 1947 Statute of the International Law Commission, article 1(1).

[112] E.A. Oji, *op cit*.

[113] 1945 Statute of the International Court of Justice.

[114] I. Brownlie, *op cit*; M.O. Hudson, *op cit*. According to Professor Shaw, *"In international law, it is a dynamic source of law in the light of the nature of the international system and its lack of centralized organs"* – M.N. Shaw, *op cit*, p. 70.

[115] *Ibid*, p. 73.

However, in the process of such logical deduction, a distinction has consistently been drawn between custom strictly so-called and usage. Usage represents an international habit of action without legal obligations, whereas custom represents those usages which have obtained the force of law.

It is instructive to note that the very nature of customary international law crystallizes as a necessarily incidental precipitate of the distinction above stated and may be summarily discussed under the following heads, *viz*: actual behavior of states (i.e. usage) including the elements of duration, generality, uniformity and consistency in such practice, and the psychological or subjective belief that such behavior cum usage is 'law'.

Actual Behaviour of States

Actual behavior of states means such usage as appears from the practice of states, which practice may take several forms such as treaties, decisions of international and municipal courts, municipal legislation, diplomatic correspondence, opinions of national legal advisers, practice of international and national institutions, policy statements, official manuals on legal questions (e.g. manual of military law), as well as executive decisions and practices.[116] According to Professor Shaw, "it is understandable why this first requirement…, since customary law is founded upon the performance of state activities and the convergence of practices, in other words, what states actually do".[117]

Although custom develops from recurrent acts of state practice, no particular duration is required in international law such that a lot will depend on the circumstances of the case and the nature of the usage in question.[118] Duration is thus not the most important component of state practice, though the practice must have gone over a period during which it becomes obvious that it is general in nature.[119]

The element of generality of practice means that a large number of states must

---

[116] *Arrest Warrant Case (Congo* v *Belgium)*, ICJ Reports, 2002, p. 3 at 23 – 24; *Interhandel Case (Switzerland* v *United States of America)*, ICJ Reports, 1959, p. 27; *Reparation Case*, ICJ Reports, 1949, p. 174; *Yearbook of the ILC*, 1950, vol. II, pp. 368 – 372; J.L. Brierly, *The Law of Nations*, 6th edn, Oxford, Oxford University Press, 1963, p. 60.
[117] M.N. Shaw, *op cit*, 74.
[118] *Ibid*, p. 76.
[119] A.A. D'Amato, *Concept of Custom in International Law*, Ithaca, Cornell University Press, 1971, pp. 56 – 58; M. Akehurst, "Custom as a Source of International Law", *BYIL* (1974-1975) 47(1), pp. 15 – 16; But see the contrary opinion to the effect that immemorial usage is required – *European Commission of the Danube* Case (1927) PCIJ Series B No. 14 p. 105 *per* Judge Negulesco.

have adopted the practice. This does not mean universality in the sense that all states must adopt the practice. The position is more lucidly espoused by Professor Shaw as follows:

> The reason why a particular state acts in a certain way are varied but are closely allied to how it perceives its interests. This in turn depends upon the power and role of the state and its international standing. Accordingly, custom should to some extent mirror the perceptions of the majority of states, since it is based upon usages which are practiced by nations as they express their power and their hopes and fears. But it is inescapable that some states are more influential and powerful than others and that their activities should be regarded as of greater significance. This is reflected in international law so that custom may be created by a few states, provided those states are intimately connected with the issue at hand, whether because of their wealth and power or because of their special relationship with the subject-matter of the practice, as for example maritime nations and sea law. Law cannot be divorced from politics or power and this is one instance of that proposition.[120]

The rule relating to uniformity and consistency of practice was laid down in the *Asylum Case*[121] where the International Court of Justice declared that a customary rule must be "in accordance with a constant and uniform usage practiced by the states in question".[122] The need for uniformity and consistency of practice does not require complete uniformity, but there should be substantial uniformity. In the *Anglo-Norwegian Fisheries Case*[123] the court refused to accept the existence of a ten-mile rule for bays because there was no uniform practice in this respect.

b. The Psychological or Subjective Element

This element, otherwise referred to in legal terminology as *opinio juris necessitatis,* was first formulated by the French writer Francois Geny as an

---

[120] M.N. Shaw, *op cit,* p. 79. Also see the *North Sea Continental Shelf* Cases, ICJ Reports, 1969, p. 3 at 42 – 43.
[121] *Asylum Case (Colombia* v *Peru)*, ICJ Reports, 1950, p. 266.
[122] *Ibid,* p. 276 – 277.
[123] ICJ Reports, 1951, p. 116 at 131; *Nicaragua* v *United States Case, supra*; *Right of Passage through Indian Territory* Case, ICJ Reports, 1960, p. 40; D.P. O'Connell, *International Law,* London, Stevens, 1970, pp. 15 – 16; D. Carreau, *Droit Internationale,* Paris, Pedone, 1994, p. 230.

attempt to differentiate legal custom from mere social usage. It relates to the belief by a state that behaved in a certain way that it was under a legal obligation to act that way.[124] Thus, the *opinio juris,* or belief that a state activity is obligatory, is the factor which turns a usage into a custom and renders it part of the rules of international law. Put differently, states will behave in a certain way because they are convinced that it is binding upon them to do so. Pontificating the need for this element in the emergence of a customary rule of international law, the International Court of Justice, in the *North Sea Continental Shelf* Cases, held *inter alia:*

> The states concerned must therefore feel that they are conforming to what amounts to a legal obligation. The frequency, or even habitual character of the act is not in itself enough. There are many international acts, e.g. in the field of ceremonial and protocol, which are performed almost invariably, but which are motivated by considerations of courtesy, convenience or tradition, and not by any sense of legal duty.[125]

It must however, be noted that although a customary rule may arise notwithstanding the opposition of one or a few states provided the necessary generality is reached, the rule so created will not bind such persistent objectors unless they are rules of a very fundamental nature postulating by their nature universality of application (e.g. rules connoting obligations *erga omnes)* or those partaking of the nature of *jus cogens.*[126] Such protests, when reinforced by acquiescence from other states, may create a recognized exception to the rule. The same cannot however, be said to apply to subsequent protests by a state after the full formation of a customary rule. Although the possibility of a state escaping from being bound by an already established custom through subsequent protests has been noted to exist,[127] the preponderance of opinion is that such states are bound by the rule and may at best work towards the evolution of a new rule to displace the existing one.[128] Note

---

[124] M.N. Shaw, *op cit,* p. 75.
[125] *Supra,* at p. 44; *Nicaragua* v *United States Case, supra; Lotus Case (France* v *Turkey)* (1927) PCIJ Series A No. 10 p. 18.

[126] *Anglo-Norwegian Fisheries* Case, ICJ Reports, 1951, p. 139; *Asylum Case (Colombia* v *Peru), supra.*
[127] The decision in the *Anglo-Norwegian Fisheries Case, supra* appears to suggest that when a state acts contrary to an established custom and other states prefer not to react to same in such a manner that suggest acquiescence, then that state mat be considered as not being bound by the original rule. See G.I. Tunkin, *Theory of International Law,* Cambridge, Harvard University Press, 1974, p. 129.
[128] M. Virally, "Sources of International Law" in M. Sorensen (ed), *Manual of Public*

also that although generality of practice is required for the formation of customary law, it is possible for a local or regional custom among a group of states or two states only to emerge.[129]

Basis of Obligation of Customary International Law

The basis of obligation of international custom is to be found within the various theories on the basis of obligation of international law as a complete system of law. The reason for the jurisprudential rationalization of the binding effect of international law is the apparent inconsistency between the existence of that system of law and the concept of state sovereignty. According to Brierly, "…if sovereignty means absolute power and if states are sovereign in that sense, they cannot at the same time be subject to law."[130] The question then is as to whether the sovereignty of states is reconcilable with international law or better put, as to the basis on which sovereign states feel bound by and do largely obey rules of international law.

The basis of obligation in state law differs radically from the basis of obligation in international law, which difference derives essentially from the difference in the nature of both systems of law. Municipal law is an expression or emanation of the will of the people as personified in the state[131] and consists of rules recognized and enforced by the sovereign state. Municipal law is therefore, essentially a law of subordination emanating from the will of a sovereign state (with legislative, executive and judicial powers) and addressed to subjects who are bound to obey, usually under the pain of sanction.[132]

---

[131] Thus, in the national legal order, the common will of the citizens finds its melting pot in the juristic entity of the state endowed with an institutional apparatus which lends cohesion to the society and authority cum force to its laws. This position obtains even in a military dictatorship or a totalitarian state in that upon the effective super-imposition of the will of the ruling class on the people, both wills melt into the 'state will' as epitomized by the legitimacy accorded the government. This does not necessarily mean that there is no possibility of the disintegration of 'state will'. Such disintegration will either result in the crystallization of a changed will for the same state (i.e. a change in government and its institutions, which may extend to the laws applicable in the state) or it may result in the disintegration of that state and the emergence of smaller states each possessing its own will.

[132] This aligns with the Austinain command theory of law to the effect that law is a species of command by a sovereign (a determinate political superior) to his subjects (political inferiors) who are under a duty to obey with the pain of sanction for any violation – J.S. Austin, *The Province of Jurisprudence Determined*, H.L.A. Hart (ed), London, Weidenfeld & Nicolson, 1954, p. 14.

International Law, London, Macmillan, 1968, pp. 137 – 138.
[129] *Right of Passage Case through Indian Territory Case, supra.*
[130] J.L. Brierly, *The Law of Nations*, 6th edn, New York, Oxford University Press, 1963, p. 16.

Conversely, there is the absence of a sovereign law-making authority in international law, absence of a sovereign executive authority enforcing international law and absence of a supreme tribunal with compulsory and unlimited jurisdiction. Thus, the international system is one of co-ordination in the sense that the community of states who are all sovereign, constitute at the same time the subjects of international law who are bound by the laws.[133]

There are two major schools of thought on the basis of obligation in international law, *viz*: voluntarism (sometimes called positivism or consensualism) and objectivism (of which *jusnaturalism* is a variant).

Voluntarism proceeds from the fundamental assumption that rules of law are products of the human 'will', they exist for this will and also by this will. This school of thought has been theoretically rationalized from different perspectives. According Jellinek, since no state, in its attribute of sovereignty, is subject to any other state, it is the sovereign manifestation of state will that creates law. The state does this through the faculty of self-determination whereby the state creates law for itself in both internal and external affairs, and the faculty of self limitation whereby the state subjects itself, when it thinks same expedient to its private law, to recognize the personality of foreign states and bind her own will by entering into the international system. Thus, the continued obedience of states to international law is an expression of their sovereign will.[134] Triepel, on his own, opines that the 'will' that can impose on and bind sovereign states must be a 'superior will'. Since the will of no single state imposes, it has to be the common will of states. This 'common will' comes into existence through the *'vereignbarung'* which designates 'a union of wills' in which the wills of the participating states seek the same common objective in union in contradistinction to a 'contract' where the contracting wills pursue different objectives. The *'vereignbarung'* may be expressly realized as is the case in treaties, or it may be tacitly realized as in customary international law.[135] Cavaglieri rejects the view that international law rests on an external

---

[133] C. Rousseau, *Le Droit International Public*, Paris, Sirey, 1970, p. 27.

[134] G. Jellinek, *Die Rechtliche Natur der Staatenvertrage: Ein Beitrag Zur Juristischen Construction des Volkerrechts*, Wien, A. Holder, 1880, p. 10.

[135] H. Triepel and R. Brunet, *Droit International et Droit Interne*, Paris, A. Pedone, 1920, p. 15.

command and instead maintains that international law is merely a system of promises between co-ordinated and juridically equal subjects. He ascribes the juridical basis of the binding common will of states to the principle of *pacta sunt servanda*.[136] Anzilloti supports this view and maintains that the principle is an absolute postulate of the international legal system imposing on, and independently of, the will of states. Thus, it is an *a priori* assumption of the international system which itself cannot be proved juridically.[137]

Conversely, the objectivist doctrine situates the origin of the binding force of international law outside the human 'will' and places it either in a fundamental norm from where all rules emanate (such as the normativist theory of the Vienna school of thought led by Kelsen and Verdross) or in social necessities (as in the theories of Duguit and Scelle). Kelsen explains the binding force of international law on the basis of the law of normativity in which he ascribes to the principle of *pacta sunt servanda* the role of a fundamental norm which confers validity on all subordinate norms in the international legal system. In this hierarchically ordered system, each norm derives validity from a higher norm culminating in a legal pyramid. This is similar to the national legal order where the constitution, imparts validity on other statutes which in turn validates by-laws and delegated legislations culminating in the individualization of the norm by the execution of an act by an official of the system. According to him:

> We have to start from the lowest norm within international law, that is, from the decision of an international court. If we ask why the norm created by the decision is valid, the answer is furnished by the international treaty in accordance with which the court was instituted. If again we ask why this treaty is valid, we are led back to the general norm which obligates the states to behave in conformity with the treaties they have concluded, a norm commonly expressed by the phrase *pacta sunt servanda*. This is a norm of general international law, and general international law is created by custom constituted by the act of states. The basic norm of international law, therefore, must be a norm which countenances

---

[136] A. Cavaglieri, *Lezione di Diritto Internationale: Parte Generale*, Napoli, Gennato Maio, 1925, p. 44.

[137] D. Anzilotti, *Corso di Diritto Internazionale*, Paris, Sirey, 1929, p. 46.

custom as a norm-creating fact, and might be formulated as follows: 'the states ought to behave as they have customarily behaved'.[138]

Professor Brierly reminds us that there need not be any mystery about the source of the obligation to obey international law and declares that a mere juridical explanation cannot suffice to solve the problem of the obligation to obey the law. The answer must be sought outside the law. For him, the obligation to obey international law has a moral foundation as dictated by human rationality and social necessity. Thus, he declares that:

> The ultimate explanation of the binding force of all law is that man, whether he is a single individual, or whether he is associated with other men in a state, is constrained in so far as he is a reasonable being, to believe that order and not chaos is the governing principle of the world in which he has to live.[139]

For sociological jurists, the centre of gravity of law lies in the society itself. Social necessities provide not only the origin but also the basis and the validating criterion of law. Thus, Professor Leon Duguit is recounted as having opined that all laws, including international law, are products of social solidarity. The transformation of a social norm into a juridical norm (otherwise called objective law) occurs when the bulk of the members of the society accept as legitimate its regular enforcement by those in power. Adopting this logic, Scelle declares that the respect for social solidarity is not only the basis of law but is also a biological necessity since no one can compromise it without harming societal life and his own life.[140]

There is some merit in all these theories; their weakness lies in their claim to universality which taints their credibility. Indeed, it is submitted that the true basis of obligation of international law, more especially custom, is to be sought in a hybrid of all these theories and may be stated thus: 'the consent or will of the states to be bound by their common acts as necessitated by such moral and social considerations as are prevalent in the society at the relevant era'.

---

[138] H. Kelsen, *General Theory of Law and State*, Cambridge, Harvard University Press, 1945, pp. 369 – 370.

[139] J.L. Brierly, *op cit*, pp. 54 - 56.

[140] G. Scelle, *Manuel de Droit International Public*, Paris, Domat-Montchrestein, 1948, p. 6.

Relationship between International Law and Municipal Law

There are two major conceptions of the relationship between international law and municipal law which conceptions logically derive from and reflect the position adopted by theorists with relation to the basis of obligation in international law. Thus, while the voluntarist theory which ascribes the basis of obligation to the consent of states leads to 'dualism', the objectivist theory which situates the basis of obligation of law outside the human or state will favour 'monism'.[141]

Dualists' doctrines postulate that municipal law and international law constitute two distinct and separate categories of legal systems. Thus, the validity of municipal law is not conditioned by international law, such that within a state, the rules of international law cannot be applied as such, but only after being transformed or received into that legal system.[142]

The monists on the other hand, maintain that international law and municipal law must be regarded as manifestations of a single conception of law. The main reasons for this assertion is that both laws are addressed ultimately to the conduct of the same subjects (i.e. the individual) and some of the fundamental notions of international law cannot be comprehended without the assumption of a superior legal order from which the various systems of municipal law are, in a sense, derived by way of delegation.[143]

International practice does not endorse any of the competing theories of Monism or Dualism unreservedly.[144] As regards the question of primacy of international law, international jurisprudence leans in favour of monism with primacy of international law. It is, in the words of Hersch Lauterpacht, "a critical and realistic monism, fully alive to the realities of international life." He gives his reasons for this cautious view:

> Just as international law is at present an imperfect law in a stage of transition to true law; so its monistic structure is not absolute and thorough going. It is a monism qualified by dualistic

---

[141] A third theory, nihilism, preaches absolute supremacy of municipal law over international law. This theory appeared under the favourable conditions created by German militarism and was called to serve its predatory interests. See E.A. Oji, *op cit*, p. 156.

[142] E.A. Oji, *op cit*; H. Triepel, *Volkerrecht und Landesrecht*, Berlin, Leipzig, C.L. Hirschfeld, 1899, p. 7; K. Strupp, "Les Regles Generales du Droit International de la Paix", *HR* (1934)47, p. 43; M.N. Shaw, *op cit*, p. 131.

[143] E.A. Oji, *op cit*; H. Kelsen, *op cit*, p. 363 – 380; M.N. Shaw, *op cit*, p. 132.

[144] E.A. Oji, *op cit*.

exceptions and contradictions. This statement may appear paradoxical seeing that in pure juridical logic there is no transition between monism and dualism. But the very imperfection of international law implies that, if we are to give a true picture of its present position, we cannot treat it as a logical system. It is therefore necessary to admit that, so far as positive law is concerned, monism, while providing a working instrument of scientific knowledge for international law as a whole and while providing an adequate and the only possible basis for its development to true law, often breaks down and yields to the reality of a dualistic nature.[145]

Dr. Oji opines that Lauterpacht's view accords with contemporary reality. Citing the decision in the *Alabama Claims Arbitration*,[146] she pontificates that monism will ensure the survival of international law since the logic of dualism would not only be a subversion, but also a negation, of international law. In line with Dr. Oji's position, it is settled law that a state cannot plead the provisions of its own law or deficiencies in that law in answer to a claim against it for breach of an obligation under international law.[147] The principle of primacy of international law over municipal law was reaffirmed by the ICJ in its Advisory opinion in the *United Nations Headquarters Agreement* Case.[148] This principle of primacy of international law over municipal law before international tribunals applies to all aspects of a state's municipal law, to its constitutional provisions, its ordinary legislation, the executive acts of its officials and to the decisions of its courts.[149] Today, international human rights courts often declare national laws incompatible with international rules and may award

---

[145] H. Lauterpacht, *International Law, Collected Papers*, Vol. 1, Cambridge, Cambridge University Press, 1970, p. 214.

[146] (1872) Moore Arbitration, p. 653.

[147] Also see the *Free Zones* Case PCIJ Reports, Series A/B, No. 46, p. 47; the *Graeco-Bulgarian Communities* Case (1930) PCIJ Reports, Series B, No. 17, p. 32 and the *Polish nationals in Danzig* Case (1932) PCIJ Reports, Series A/B, No. 44, p. 24.

[148] ICJ Advisory Opinion of 28th April 1988.

[149] *Massey Claim* Case, 4 RIAA 155 (1927); *Chorzow Factory* Case, *supra; Peter Pazmany University* Case, PCIJ Decision of 15th December 1933; *Youman's Case (United States v Mexico)*, 4 RIAA 110 (1926); *Caire's Claim (France v Mexico)*, 5 RIAA 516 (1929); *Chattin's Claim*, 4 RIAA 282 (1927); *Exchange of Greek and Turkish Populations* Case, PCIJ Reports, Series B, No. 10, p. 20; *Finnish Ships Arbitration (Finland v UK)*, 3 RIAA 1479 (1934).

compensation to those whose rights have been violated.[150]

Application of International Law in Municipal Courts

It must be stated that in exercise of sovereignty, it is within the exclusive jurisdiction of a state to determine what laws should operate within its legal system. Thus, although a state bears an obligation to act in conformity with international law and will bear responsibility for breaches of it in the international sphere, conflict between a state's municipal law and its international obligations does not affect the effectiveness of that municipal law within the territory of that state.[151]

Perhaps, this explains why Potter, P. in the case of *Wilkinson* v *Kissinger*[152] refused to be bound by the decision of the European Court of Human Rights (ECtHR) in *Christine Goodwin* v *UK*.[153] It is therefore not surprising that there is no consistent or general behaviour by states as regards the application of international law within their municipal legal system such that the practice varies from state to state.

The U.S. Constitution designates ratified treaties, along with the Constitution itself and federal statutes, the supreme law of the land[154] and empowers Congress "to define and punish ... offences against the Law of Nations."[155] Customary international law is automatically incorporated into the U.S. legal system as federal common or unwritten law.[156] The U.S. state and federal courts presume that U.S. law conforms to international law; such an attitude has been urged consistently by the Supreme Court of the United States.[157]

The practice in the United Kingdom allows for the applicability of

---

[150] For instance, in its operation, the European Court of Human Rights may hold a state law invalid if it is against the Community law. See the case of *Christine Goodwin* v *UK* (2002) ECHR 588; E.A. Oji, *op cit*.

[151] The obligation to obey international law and the concomitant responsibility attendant to breach of same usually compels the state to take cognizance of their international obligations in the course of municipal exercise of executive, legislative and judicial powers of government. Again, the development of international law, especially international custom also takes the municipal behavior of states into cognizance. In view of the above, Professor Shaw argues that "there is indeed a clear trend towards the increasing penetration of international legal rules within domestic systems coupled with the exercise of an ever-wider jurisdiction with regard to matters having an international dimension by domestic courts. This has led to a ...greater preparedness by domestic tribunals to analyze the actions of their governments in the light of international law" – M.N. Shaw, *op cit*, p. 138.

[152] (2006) EWHC 2022 (Fam), (2006) H.R.L.R. 36.
[153] *Supra*.
[154] Article VI.
[155] Article I, Section 8.
[156] *The Paquatte Habana* (1900) US 677 20 Sup. Ct. Rep. 290.
[157] *Filartiga* v *Pena-Irala* (1980) 630 F. 2d 879.

international law[158] on the basis of the doctrine of transformation[159] and the doctrine of incorporation. The doctrine of transformation maintains that before any principle of international law can be applied in English courts, it has to be transformed or specifically adopted into English law by the use of appropriate constitutional machinery, i.e. by an Act of parliament, authoritative judicial decision or established usage.[160] Conversely, the doctrine of incorporation holds that rules of international law are automatically part of English law and are applicable in British courts provided they are not inconsistent with Acts of parliament or prior authoritative judicial decisions.[161] The modern practice in UK shows a preference for the incorporation doctrine.[162]

Considering the above, it may be asserted that while in some countries a treaty or customary international law is given constitutional status superior to national legislation,[163] in other countries treaties do not become effective in national law until they are enacted by Parliament. This latter attitude is adopted in most English

---

[158] See Upjohn J. in *Re Claim by Herbert Wragg & Co. Ltd* (1956) Ch 323 at 334; and Lord Cross in *Oppenheimer* v *Cattermole* (1976) AC 249 at 277.

[159] Otherwise called the specific-adoption theory.

[160] *R* v *Keyn* (1876)2 ExD 63; *Mortensen* v *Peters* (1906)8 F (J) 93; M.N. Shaw, *op cit*, p. 139.

[161] While expounding this theory, Blackstone opined that "the law of nations, wherever any question arises which is properly the object of its jurisdiction, is here adopted in its full extent by the common law, and it is held to be part of the law of the land". See M.N. Shaw, *op cit*, p. 140; *West Rand Gold Mining Co.* v *R* (1905)2 KB 391; *Chung Chi Cheung* v *R* (1939) AC 160; *Buvot* v *Barbuit* (1737) Cases t. Talbot 281; *Triquet* v *Bath* (1764)3 Burr 1478.

[162] *Ibid*; *Trendtex Trading Corp.* v *Central Bank of Nigeria* (1977)2 WLR 356; *Thai-Europe Tapioca Services Ltd* v *Government of Pakistan* (1975)3 All ER 961; *Maclaine Watson* v *Department of Trade and Industry* (1988)3 WLR 1033 where Nourse, LJ emphasized that the *Trendtex Case* had resolved the rivalry between the incorporation and transformation doctrines in favour of the former. Also see the dictum of Lord Slynn in *Ex Parte Pinochet (No. 1)* (2000)1 AC 61 at 77. However, the qualifications to this rule must be noted. Firstly, treaties ratified by UK except those relating to the conduct of war and concession of territory, and also offences under international law do not automatically become incorporated into English law until specifically adopted by an Act of parliament – *Maclaine Watson's Case, supra; the Parlement Belge* (1879)4 PD 129. Secondly, an Act of parliament or authoritative judicial decision prevails over any rule of international law to the contrary – *Mortensen* v *Peters, supra*. However, note the English law presumption that parliament does not intend to act in breach of international law such that an Act of parliament should be interpreted in a manner as would avoid conflict with international law – *Garland* v *British Rail Engineering Ltd* (1983)2 AC 751; *Ex Parte Brind* (1991)1 AC 696. Thirdly, on issues relating to the status of a foreign state or government, or the existence of a state of war, a certificate signed by the Foreign Secretary is conclusive of the issue and overrides any position adopted on the issue by international law – *The Annette* (1919) P 105; *City of Berne* v *The Bank of England* (1804)9 Ves. Jun. 347. Fourthly, by virtue of the act of state doctrine under English law, an alien injured abroad by an act authorized or ratified by the Crown has no remedy in English courts, despite any rule of international law to the contrary – *Buttes Gas & Oil Co.* v *Hammer (No. 3)* (1982) AC 888; *Buck* v *Attorney General* (1965)1 Ch 745; *Helen Liu* v *Republic of China*, 29 ILM, 1990, p. 192.

[163] See the Basic Law of the Federal Republic of Germany, Article 25; Dutch Constitution, Article 65; 1947 Italian Constitution, Article 10.

speaking countries of Africa. Most of such countries require an Act of Parliament to incorporate international law into municipal law before it can be enforceable.[164] Most of the Constitutions in their provision on the applicability of international law within the courts of the state refer to treaty law. Nothing is said on the status of international customary law before these courts. Only one African Constitution,[165] that of South Africa, in 1994 and 1996 explicitly refers to customary law. According to section 231, "the rules of customary international law binding upon the Republic shall … form part of the law of the Republic". The implication of the above situation is lucidly captured by Dr. Oji in the following words:

> In consequence, it would appear that international customary law only becomes incorporated on the basis of the acceptance of states to act in accordance with the general rules of international law. What is the implication of such a situation? When it is noted that treaties which have the positive consent of states in signature and ratification are mostly subjected to parliamentary re-enactment and or acceptance, it only leaves to imagination what may be the attitude of some of the states whose Constitution is silent on the status of international customary law. Especially, realizing how international customary law is established, it may give room to defaulting states to argue against its positive nature.[166]

Though section 12 of the Nigerian Constitution provides only for the applicability of treaties ratified by the country thus suggesting the municipal non-enforceability of international customs,[167] an analysis of the Nigerian legal system reveals the position of customary international law in Nigeria.

Nature of the Nigerian Legal System

The Nigerian legal system has a chequered history.[168] From the pre-

---

[164] See section 1999 CFRN (as amended), section 12; Constitution of South Africa, Article 242; Namibia requires that the parliament does not object to the international law for it to be effective.

[165] M.T. Ladan, *Materials and Cases on Public International Law*, Zaria, Ahmadu Bello University Press Ltd, 2007, p. 6.

[166] E.A. Oji, *op cit*, pp. 161 – 162.

[167] This absurd and totally unwelcome proposition appears to be reinforced by the '*expressio unius est exclusio alterius*' maxim of statutory interpretation (which literally means that the express mention of a thing excludes all others) – *Peoples Democratic Party* v *INEC* (2001)1 WRN 1; *Richardson* v *Lead Smelting Co.* (1762)3 Burr 1341.

[168] An elaborate discussion on the history of the Nigerian legal system may be found in C.J.S. Azoro, "The Place of Morality in the Nigerian Legal System: A Jurisprudential Approach", an

colonization era when all the different ethnic groups that comprise the country each had its own set of rules and practices governing life in their respective societies and also various institutional frameworks for the administration and enforcement of these rules, the incursion of colonization introduced a radical change in the nature of the Nigerian legal system. It brought about the introduction of English law (both in its received and extended form) and also the establishment of English-styled courts in Nigeria. The post-independence era produced the legal system as we have it today with the Constitution providing for the making of the various laws and the establishment of the various institutions regulating affairs in the country. The sources of Nigerian law currently include the Constitution, Nigerian legislation, Nigerian case-law, customary law, English law and international law.

The Nigerian Constitution is the fundamental law of the land and specifies a bundle of rights and duties, as well as rules that may be enforced under the law.[169] One remarkable feature of all Nigerian constitutions is that they have all been written. Unlike the position in Britain where parliament is supreme,[170] the Nigerian Constitution is superior to all other laws of the land and regulates the judicial, executive and legislative organs of government as well as the rights of the citizens.[171] Thus, it is the basic norm; the ultimate premise of the legal system.[172]

Nigerian legislation consists of all Acts, Laws, and subsidiary legislations in force in Nigeria. All enactments made by the National Assembly are designated as 'Acts' while those made by the Houses of Assembly of the various states are designated as 'Laws'.[173]

---

unpublished LL.B Project submitted to the Faculty of Law, Nnamdi Azikiwe University, Awka, pp. 37 – 42.

[169] For example, Chapter IV which provides for fundamental rights and freedoms.

[170] *Lee* v *Bude & Torrington Railway* (1871) LR 6 CP 576; *R* v *Jordan* (1967) Crim LR 483; *Chenny* v *Conn* (1968)1 WLR 242.

[171] 1999 CFRN, section 1(1) & (3); *INEC* v *Musa* (2003)10 WRN 1; *Attorney General of Abia State* v *Attorney General of the Federation* (2002)17 WRN 1, [2002]6 NWLR (pt 763) 264; *Attorney General of Ondo State* v *Attorney General of the Federation* (2002)27 WRN 1, [2002]9 NWLR (pt 772) 222.

[172] D Lloyds, *The Idea of Law*, London, Penguin Books, 1979, p. 194.

[173] When federalism was introduced in Nigeria in 1954, all enactments made by the central legislature prior to 1st October 1954 retained the name 'Ordinances' while those of the regional legislatures were designated 'Laws'. On attainment of independence in 1960, the laws made by the federal legislature were renamed 'Acts' while those of the Regions continued to be 'Laws'. Upon military intervention in Nigerian political life, enactments made by the Federal Military Government became known as 'Decrees' while laws made by the State Military Governors or Administrators were known as 'Edicts'. With the return of democratic government, the 'Decrees' were renamed 'Acts' while the 'Edicts' were renamed 'Laws'.

Though the traditional role of the courts is *jus dicere* and not *jus dare*,[174] the Nigerian judges still make law, albeit in a different sense from the legislature.[175] For instance, where there is no law previously governing the situation before the court, the judge may create some principles of law for the situation.[176] Furthermore, the judges have to apply the law to ever-changing combinations of circumstances to which the law has never been previously applied. Where a court declares a rule for purposes of deciding a case, such rule becomes a precedent for deciding future cases with similar facts. Judicial precedents is therefore one of the sources of Nigerian law.[177]

Customary law in Nigeria is traditionally classified into ethnic/non-muslim law and sharia law. The ethnic/non-muslim law consists of the various indigenous laws applicable to the different ethnic groups in Nigeria. Islamic law applies to adherents of that religion and was introduced into Nigeria as an aftermath of the successful process of Islamization and the *jihads* in Northern Nigeria. It is based on the Holy Koran and the teachings of the Prophet Mohammed as interpreted by the rightly guided Caliphs. In some areas, Islamic law has completely supplanted the pre-existing customary laws, while in others, there has been a relative fusion of the two systems. The teachings of the Maliki school of thought is predominantly applied in Nigeria.[178] Section 14 of the Evidence Act makes provisions for the application of customary law in Nigeria.[179]

English law (in its received and extended forms) are also part of Nigerian law as an incidence of colonialism. The extended English law refers to those English statutes made by the Crown and her agents, which were made to apply directly to Nigeria.[180] Received English

---

[174] i.e. to state and declare the law, not to give law. See the 1999 CFRN (as amended), section 6; *Attorney General of Ondo State v Attorney General of the Federation*, supra; *Lakanmi v Attorney General of Western Nigeria*, supra; *Attorney General of Abia State v Attorney General of the Federation* (2006) NSCQR 161; *Attorney General of Ogun State v Attorney General of the Federation* (1982)1-2 SC 13; *Attorney General of Lagos State v Attorney General of the Federation* (2004)20 NSCQR 29; P.A.O. Oluyede, *Constitutional Law in Nigeria*, Ibadan, Evans Bros Publishers Ltd, 1992, pp. 75 – 78.

[175] *Ogunlowo v Ogundare* [1993]7 NWLR (pt 307) 610; C.K. Allen, *Law in the Making*, 7th edn, Oxford, Clarendon Press, 1964, p. 16.

[176] *Bello v Attorney General of Oyo State* (1986)12 SC 1; *Obi v INEC* [2007]11 NWLR (pt 1046) 565; *Amaechi v INEC* [2007]18 NWLR (pt 1065) 2, (2007)7-10 SC 172.

[177] J.O. Asein, *Introduction to Nigerian Legal System*, 2nd edn, Lagos, Ababa Press Ltd, 2005, pp. 73 – 97; O.N. Ogbu, *Modern Nigerian Legal System*, 2nd edn, Enugu, CIDJAP Press.

[178] J.O. Asein, *op cit*, p. 118.

[179] Evidence Act, 2011; *Agbai v Okogbue* [1991]7 NWLR (pt 204) 391; *Oyewunmi v Ogunesan* [1990]3 NWLR (pt 137) 137; *Ojisua v Aiyebelehin* [2003]11 NWLR (pt 723) 44.

[180] The Foreign Jurisdiction Acts (UK), 1843-

law refers to the principles of common law, doctrines of equity and statutes of general application in force as at 1st January 1900 which were incorporated into Nigerian law by local legislations.[181]

International law is one of the sources of Nigerian law. Albeit at risk of prolixity, it must be emphasized that the Nigerian Constitution provides for the domestic application of any treaty ratified by Nigeria provided it has been transformed into Nigerian law by an Act of the National Assembly.[182] As will be seen shortly, international customs also form part of Nigerian law.

Customary International Law and the Nigerian Legal System

It is quite unfortunate that the position of customary international law in Nigeria is as clear as mud.[183] Section 12 of the Nigerian Constitution provides only for the applicability of treaties ratified by the country thus suggesting the municipal non-enforceability of international customs. This absurd and totally unwelcome proposition appears to be reinforced by the *'expressio unius est exclusio alterius'* maxim of statutory interpretation (which literally means that the express mention of a thing excludes all others).[184]

Though section 19(d) of the Nigerian Constitution provides that the foreign policy objectives of the country shall be the respect for international law and treaty obligations as well as the seeking

---

1913 and the Colonial Laws Validity Act (UK), 1865 gave this power to the Crown. Since independence, all Nigerian Constitutions have preserved these laws. See 1960 Nigerian Constitution (Order-in-Council), section 3(1); 1963 CFRN; 1979 CFRN, section 274 and the 1999 CFRN (as amended), section 315. Also see the case of *Ibidapo* v *Lufthansa Airlines* [1997]4 NWLR (pt 498) 124 where the Supreme Court held that from 1960 till date, all the English laws, multilateral and bilateral agreements concluded and extended to Nigeria, unless expressly repealed or declared invalid by a court of law or tribunal established by law, remained in force subject to the provisions of the prevailing Nigerian Constitution.

[181] Interpretation Act, Cap I23, *Laws of the Federation of Nigeria*, 2004, section 32; High Court Law (Eastern Nigeria), Cap 60, *Laws of Eastern Nigeria*, 1963, section 3; Law of England (Application) Law (Western Nigeria), Cap 60, *Laws of Western Nigeria*, 1959, sections 28 and 29; High Court Law (Northern Nigeria), Cap 49, *Laws of Northern Nigeria*, 1963, section 35. Note that with the successive creation of states from the different regions, the new states adopt the laws of the parent region, sometimes with minor amendments. The received English law reception clauses of the relevant enactments in almost all the states are similar. Note also that the reference date of 1st January 1900 has been held to apply only to statutes of general application so as to allow for the application of the principles of common law and doctrines of equity in their dynamic nature and as perceived by the English courts from time to time. See the case of *Nigerian Tobacco Co. Ltd* v *Agunanne* (1995) LPELR-SC.31/1989, [1995]5 NWLR (pt 397) 541. Also see J.O. Asein, *op cit*, p. 107.

---

[182] *General Sani Abacha* v *Gani Fawahinmi*, supra.
[183] This is unlike the position in Ghana as seen in the preceding chapter of this work.
[184] *Peoples Democratic Party* v *INEC* (2001)1 WRN 1; *Richardson* v *Lead Smelting Co.* (1762)3 Burr 1341.

of settlement of international dispute by negotiation, mediation, conciliation, arbitration and adjudication, this section is not enough to warrant the application of customary international law in Nigeria. This assertion is predicated on the fact that the section is dedicated to the foreign policy objectives which Nigeria as a state pursues and nothing more. Besides, section 6(c) of the Constitution makes, not only the provisions of section 19, but also the provisions of the entire Cap. II of the Nigerian Constitution non-justiceable.

The obvious lacuna in the Constitution is capable of keeping one in the dark as regards the applicability or otherwise of customary international law. There is a dearth of Nigerian judicial authorities on the issue. In the case of *Ibidapo v Lufthansa Airlines*,[185] the Supreme Court failed to advert its mind to this issue while pronouncing on the position of international law in Nigeria and only focused on bilateral and multilateral agreements. The only Nigerian case that dealt with customary international law is the case of *African Continental Bank* v *Eagles Super Pack Ltd*.[186] In that case, the issue for determination was whether the Uniform Customs and Practice (UCP) for documentary credit is applicable in Nigeria. The UCP was made by the International Chambers of Commerce with headquarters in Paris with a view of having a universal standardization of letters of credit in banking and commercial transactions. At the trial court, it was held, *per* Ononuju J. that the UCP is not applicable in Nigeria. However, at the Court of Appeal, it was held that the UCP constitutes customary international law and can be judicially noticed and applied in Nigeria. Indeed, the Supreme Court, in *Akinsanya* v *United Bank for Africa*,[187] applied the provisions of the UCP although it was neither argued nor decided that it amounts to an international custom and whether same is applicable in Nigeria by virtue of that.

From the decision of the Court of Appeal in *African Continental Bank* v *Eagles Super Pack Ltd.*, and the attitude of the apex court in *Akinsanya* v *United Bank for Africa*, it may be argued that Nigerian courts can judicially notice an international custom under the provisions of the Evidence Act.[188] However, the validity of such an argument is doubtful considering the fact that section 258 of

---

[185] *Supra.*
[186] [1995] 2 NWLR (pt 379) 590
[187] [1986] 4 NWLR (pt 35) 273
[188] Evidence Act 2011, s.17 (formerly Evidence Act, Cap E14, *Laws of the Federation of Nigeria*, 2004, section 14, which provides for judicial notice of custom).

Evidence Act[189] defines custom to mean 'a rule which in a particular district has from long usage obtained the force of law'. The word 'district' has been defined to mean 'area of a country or town especially one that has a particular feature'.[190] Juxtaposing these two definitions, it can be seen that for a custom to be susceptible to the invocation of the principle of judicial notice under the Nigerian Evidence Act, such a custom must be that of a locality in Nigeria. The draftsman never intended any custom outside Nigeria, such that the attitude of the courts as depicted in the cases above remain of doubtful validity.

From the above, it becomes clear that the fate of customary international law as regards its applicability in Nigeria remains marred by uncertainties. Dr. Oji seriously criticizes the current position and argues for the application of international customs in Nigeria.[191] After a critical analysis of the nature of Nigerian customary law and an analogous exposition on the similarities between the two systems of law, she makes a case for the application of international customs in Nigeria, just on the same terms as Nigerian customary law. According to her:

> ...if ethnic customary law can form part of the body of Nigerian laws, so also can international customary law. It may only require that such international customary law be established before the Nigerian court; and that it passes the repugnancy test; incompatibility test and the public policy test.[192]

She pontificates that the requirement of passing the repugnancy test will not constitute any problem, as before any norm of international practice can translate into international customary law, it would have passed a stiffer test, that is, acceptability by a large number of the international community. Any practice that acquires such a generality of acceptance would certainly not be repugnant to natural justice, equity and good conscience. The requirement itself is such that is accepted by most civilized nations.[193]

---

[189] *Ibid*. This section is *ipsisima verba* with section 2 of the repealed Act.
[190] A.S. Hornby, *Oxford Advanced Learners' Dictionary of Current English*, 7th edn, Oxford, Oxford University Press, 2006) p.426.
[191] E.A. Oji, *op cit*, pp. 163 – 167.
[192] *Ibid*, pp. 164 – 165.
[193] The ICJ has increasingly referred to "equity" in its judgments. For example, in the *Guld of Maine* case, (ICJ Report (1984) 246 at 305, it stated that the concepts of acquiescence and estoppel in international law follow from the fundamental principles of good faith and equity. It also referred to considerations of equity in the *Barcelona Traction* case. See the case of where the ICJ applied the principle of equity.

She envisages a problem as regards the compatibility test since several of the new international norms seek to change the *status quo*. To solve this, she falls back on the purpose of that requirement for the validity of ethnic customary law which is to make sure that there is consistency in the existence and application of law in the country and to abolish customary laws that conflict with the provisions of the Constitution and other laws made by the Nigerian legislature. She suggests that just as with ethnic customary law, for international customary law to be enforceable within the states, it should not be incompatible with any law for the time being in force. Her reason for arguing to sustain this position is that if it is not so, the position of the law, at some times may be unascertainable, especially during the window periods of the emergence of international customary law. Again, it will also be possible for international customary law to define the rights and liabilities of citizens without any input by them through their elected representatives.[194]

As regards the public policy test, she argues that it is the public policy of Nigeria, and not that of the international community that will be relevant. This condition will take care of the customs and peculiar traditions of the country. For instance, the international policy may accept a norm that is totally alien and not in conformity with the belief of a people. For instance, the body of international human rights is growing rapidly, to protect certain minority groupings that some African culture may bluntly refuse to accord recognition. Thus, an international custom seeking to protect the rights of transsexuals may not be readily accepted in Africa, as not reflecting the immediate human rights challenges of the people. Treaty obligations consider this aspect of a people, thus the need for consent and provision for reservations in some cases.[195] From the above, it is clear that Dr. Oji does not suggest that section 17 of the Evidence Act expressly or impliedly provides for the domestic application of international customs in Nigeria. Rather, her position is that Nigerian courts can proactively invoke their judicial powers towards applying international customs on the same basis as local customs considering the analogy between both systems of laws. Laudable as the logic in the above position may seem, it is our humble submission that it raises several issues of jurisprudential relevance which cast serious doubts as to its practicability.

---

[194] E.A. Oji, *op cit*, pp. 165 – 166.

[195] *Ibid*, pp. 166 – 167.

First is the jurisprudential question of the basis upon which to found the domestic obligation to obey and apply international customs in Nigeria. As we have earlier submitted, the basis of the obligation to obey international law is the consent or will of the states to be bound by their 'common acts' as necessitated by such moral and social considerations as are prevalent in the society at the relevant era. What then will be the basis for the domestic application of international customs in Nigeria? This raises the vital issue of the public policy test by Dr. Oji. From her standpoint, the relevant 'will' is no longer the 'common will' of the various sovereign states from which the custom evolved. Rather, that 'will' is now made subject to the 'will' of the Nigerian people as deducible from the public policy of Nigeria. Since *opinio juris* is a vital element for international custom, it follows that unless Oji's 'public policy' is arrived at, international custom is inapplicable in Nigeria. It must be emphasized that 'public policy' is an ever-evolving concept and is not contained in a single document. It is not even the policy of the Judge who is to apply the international custom. It is the view of the generality of the Nigerian people on any particular issue that constitutes her public policy on an issue.

The plurality cum heterogeneity of the Nigerian socio-ethnic polity and the resultant differences in opinion on most issues will mean a difficulty in ascertaining the 'common will' of Nigerians on most subject matters of international custom and will invariably, affect its applicability.

The second is the issue of the apparent inconsistency between the logic of her position and Nigerian sovereignty. It is trite that in exercise of sovereignty, states reserve the authority to determine the laws that should operate in their legal system and for countries like Nigeria operating a written Constitution, same constitutes the *alpha* and *omega* of its legal system. This means that any law that is not expressly or impliedly allowed by the Nigerian Constitution, it forbids. This is the essence of the dictum of Niki Tobi, JCA (as he then was) in *Phoenix Motors Ltd* v *NPFMB*[196] where he observed that "the Constitution is the highest law of the land. All other laws bow or kowtow before it. No law which is inconsistent with it can survive. That law must die and for the good of the society…" Dr. Oji did not, and clearly could not have been able to expressly or impliedly trace her logic to the Nigerian Constitution or any law deriving validity

---

[196] *Supra.*

thereunder. Thus, she appears to posit that contrary to the principle of separation of powers under the Nigerian Constitution and in disregard of Nigeria's sovereignty, our Judges can make law by directly importing rules of customary international law without any constitutional backing. To that extent, we submit that the validity of her views remains in doubt.

Thirdly, she appears to have reduced international law to the same status as indigenous customary law. This is quite opposite the dictum of the Supreme Court in *Gen. Sanni Abacha* v *Gani Fawehinmi*[197] to the effect that rules of international law in their domestic application, will prevail over any local rule of law to the Contrary, subject to the provisions of the Constitution on their applicability.

Lastly, she appears to suggest that whatever is not repugnant to the international community will also pass the repugnancy test under Nigerian law since any practice that acquires such a generality of acceptance would certainly not be repugnant to natural justice, equity and good conscience. Thus, her requirement is met once such custom is accepted by most civilized nations. However, the fact that the general practice acceptable in most civilized countries allow same sex marriage and transexualism,[198] a practice considered repugnant under both Nigerian customary and statutory law casts serious doubts on the validity of her assertion.[199] The repugnancy test under section 17 of the Evidence Act is the Nigerian standard and not that of the international community.

The various pitfalls in Dr. Oji's view necessitate a reflection on the nature of the Nigerian legal system so as to deduce a better rationale for the domestic application of international custom in Nigeria. It is trite that "customary international law is part of the common law of England."[200] Also, it is trite that the common law is made part of the Nigerian legal system by section 32 of the Interpretation Act. The Interpretation Act is an Act of the National Assembly, validly made in exercise of the legislative powers conferred on that body by section

---

[197] *Supra.*

[198] *Christine Goodwin* v *UK* (2002) ECHR 588; *Niemietz* v *Germany* (1992) 16 EHRR 97; *Baerhr* v *Lewin* (1993) US 825 P 2d 44; *Re Kevin (validity of marriage of transsexual)* (2001) FamCA 1074; *MT* v *JT* (1976)355 A. 2d. 20k.

[199] Penal Code, Cap P3, *Laws of the Federation of Nigeria*, 2004, sections 284 and 405; *Okonkwo* v *Okagbue* [1994]9 NWLR (pt 368) 301; *Mogaji* v *Nigerian Army* [2008]8 NWLR (pt 1089) 338. Also, see the Same Sex Marriage (Prohibition) Act, 2013.

[200] Per Lord Millet in *Ex Parte Pinochet (No. 3), supra*. Also see *Lord Advocate's Reference No. 1 of 2000* (2001) SLT 507 at 512; *R* v *Jones* (2006) UKHL 16; *Commercial and Estates Co. Of Egypt* v *Board of Trade* (1925)1 KB 271.

4 of the Constitution.[201] A logical juxtaposition of the above position clearly reveals that customary international law is part of Nigerian law, applicable by our courts to the same extent as the common law. It is therefore our contention that just as is the position in England following the theory of incorporation, customary international law is also part of Nigerian law provided it is not inconsistent with the Constitution or any local enactment, or any authoritative decision of our courts. The relationship between it and customary law is to be determined on the principles of internal conflict of law.[202]

It is our submission that this view takes care of the various pitfalls that inundate Dr. Oji's position. Firstly, the basis of obligation will still remain the 'common will' of Nigeria as a state as reflected in her Constitution. This preserves the sovereignty of Nigeria and ensures the supremacy of her Constitution, since the application derives from the legislative powers provided for by the Constitution. It also avoids the problems associated with the repugnancy and public policy tests since the two tests are not relevant considerations for the application of common law in Nigeria.

Conclusion

Considering the role of international law in the maintenance of world peace and the realization of the common ideals of mankind, the importance of its application even in the municipal level cannot be over-emphasized. In the international level, this has led to the increased adoption of treaties and the proliferation of international institutions aimed at boosting greater participation in the development and enforcement of international law. However, the fact that a state may refuse to ratify a treaty and for those that apply the transformation doctrine, refuse to domesticate an already-ratified treaty poses a great threat to the realization of the ideals intended by the founding fathers of international law. Nigeria is a typical example, as the provisions of section 12 of her Constitution has denied domestic potency to the numerous treaties she has ratified, amongst which is the 1979 Convention on the Elimination of all Forms of Discrimination against Women (CEDAW).

---

[201] *Attorney General of the Federation v Guardian Newspapers Ltd* (1999)9 NWLR (pt 618) 196; *Attorney General of Abia State v Attorney General of the Federation* [2002]6 NWLR (pt 763) 300 SC.

[202] *Labinjoh v Abake* (1924)5 NLR 33; *Okolie v Ibo* (1958) NRNLR 89; *Griffin v Talabi* (1948)12 WACA 371; *Nelsen v Nelsen* (1951)13 WACA 248; *Salau v Aderibigbe* (1963) WNLR 80; *Koney v Union Trading Co.* (1934)2 WACA 188; *Osuro v Anjorin* (1946)18 NLR 45.

The nature of customary international law enables it to escape these impediments to the application of treaties above stated, since all states are bound by same, subject to few exceptions. It does so by committing them to uphold certain principles that comprise the "laws of nations" or "the customs of nations", an indication of social contract obligations on the international level. It most times imposes *erga omnes* obligations on the states to enforce its principles.[203] Apart from the fact of its enforceability against the state at the international level, all that is required for the domestic application of international customs is the appropriate constitutional machinery. In the Nigerian context, this is provided for by section 4 of the Constitution and section 32 of the Interpretation Act.

In view of the foregoing, a call is therefore made on the various institutions and agencies exercising governmental power in Nigeria to become alive to the potency and applicability of this branch of international law within the Nigerian legal system. The judiciary, as the last hope of the common man, is hereby also urged to apply the principles of common law wherever necessary to meet the justice of the numerous cases that are litigated before them, especially in those areas of Nigerian law that are yet undeveloped. It therefore behooves Nigeria, as a sign of credible commitment to her international obligations, to strive to apply international customary law towards fulfilling her pledge to the international community.

---

[203] These norms are also referred to as '*jus cogens*'. See E.A. Oji, *op cit*, pp. 168 – 169.

References

1. E.A. Oji, "Application of Customary International Law in Nigerian Courts", *NIALS Law and Development Journal* (2011) 1(1), p 151.
2. J.A. Dada, "Human Rights under the Nigerian Constitution: Issues and Problems", *International Journal of Humanities and Social Science* [Special Issue - June 2012]2 (12), pp. 33 – 43.
3. M.N. Shaw, *International Law,* 6th edn, New Delhi, Cambridge University Press, 2008, p. 70.
4. Henkin et al, *International Law: Cases and Materials,* St. Paul-Minnesota, West Publishing Co., 1980, p. 73.
5. 1969 Vienna Convention on the Law of Treaties, article 26; *Gabcikovo-Nagymaros Project Case (Hungary* v *Slovakia)*, ICJ Reports, 1997, p. 7
6. M.N. Shaw, *op cit,* p. 98; H. Thirlway, "The Law and Procedure of the International Court of Justice", *BYIL* (1988), p. 76; P. Weil, "The Court Cannot Conclude Definitely…? *Non Liquet* Revisited", *Columbia Journal of Transitional Law* (1997)36, p. 109; E.A. Oji, *op cit,* p. 154 – 155.
7. A.A. D'Amato, *Concept of Custom in International Law,* Ithaca, Cornell University Press, 1971, pp. 56 – 58; M. Akehurst, "Custom as a Source of International Law", *BYIL* (1974-1975) 47(1), pp. 15 – 16
8. M. Virally, "Sources of International Law" in M. Sorensen (ed), *Manual of Public International Law,* London, Macmillan, 1968, pp. 137 – 138.
9. G. Jellinek, *Die Rechtliche Natur der Staatenvertrage: Ein Beitrag Zur Juristischen Construction des Volkerrechts,* Wien, A. Holder, 1880, p. 10.
10. H. Triepel and R. Brunet, *Droit International et Droit Interne,* Paris, A. Pedone, 1920, p. 15.
11. A. Cavaglieri, *Lezione di Diritto Internationale: Parte Generale,* Napoli, Gennato Maio, 1925, p. 44.
12. D. Anzilotti, *Corso di Diritto Internazionale,* Paris, Sirey, 1929, p. 46.
13. H. Kelsen, *General Theory of Law and State,* Cambridge, Harvard University Press, 1945, pp. 369 – 370.
14. G. Scelle, *Manuel de Droit International Public,* Paris, Domat-Montchrestein, 1948, p. 6.
15. E.A. Oji, *op cit;* H. Triepel, *Volkerrecht und Landesrecht,* Berlin, Leipzig, C.L. Hirschfeld, 1899, p. 7; K.
16. Strupp, "Les Regles Generales du Droit International de la Paix", *HR* (1934)47, p. 43; M.N. Shaw, *op cit,* p. 131.

17. H. Lauterpacht, *International Law, Collected Papers,* Vol. 1, Cambridge, Cambridge University Press, 1970, p. 214.

18. M.T. Ladan, *Materials and Cases on Public International Law,* Zaria, Ahmadu Bello University Press Ltd, 2007, p. 6.

19. *Lee* v *Bude & Torrington Railway* (1871) LR 6 CP 576; *R* v *Jordan* (1967) Crim LR 483; *Chenny* v *Conn* (1968)1 WLR 242.

20. *Ogunlowo* v *Ogundare* [1993]7 NWLR (pt 307) 610; C.K. Allen, *Law in the Making,* 7th edn, Oxford, Clarendon Press, 1964, p. 16.

21. *Bello* v *Attorney General of Oyo State* (1986)12 SC 1; *Obi* v *INEC* [2007]11 NWLR (pt 1046) 565; *Amaechi* v *INEC* [2007]18 NWLR (pt 1065) 2, (2007)7-10 SC 172.

22. J.O. Asein, *Introduction to Nigerian Legal System,* 2nd edn, Lagos, Ababa Press Ltd, 2005, pp. 73 – 97; O.N. Ogbu, *Modern Nigerian Legal System,* 2nd edn, Enugu, CIDJAP Press.

23. A.S. Hornby, *Oxford Advanced Learners' Dictionary of Current English,* 7th edn, Oxford, Oxford University Press, 2006) p.426

OSU CASTE: A CRITIQUE
Emmanuel Okonkwo

ABSTACT

The theme of segregation is not alien to any part of the world. No matter the appellation it is branded with, its existence cannot be denied. Once, the blacks were referred to as the 'black monkeys'. In the United States, there was a glaring distinction between the white man and the black man. In South Africa, we were plunged into the dreadful arena of the Apartheid. In Nigeria, the story is the same. One wonders, are humans not made alike? Is there any justification behind this prejudice? Why is the concept of Osu prevalent? Why is it so powerful that the elites champions or are silent to it? If it is a culture, cant there be a cultural change? Why does it still exist despite the laws made against it? This essay seeks to make clear the origin, misconceptions and criticism against the Osu caste system, using the igboland of Nigeria as a case study. The method is both historical and analytical.

Keywords:

Segregation, Osu Caste, Osu caste system, igboland of Nigeria

For Referring this Paper:

Okonkwo, Emmanuel (2014). Osu Caste: A Critique. *International Journal of Research (IJR)*. Vol-1, Issue-3. Page

Note:

A SEMINAR PRESENTATION AT THE DEPARTMENT OF PHILOSOPHY, FACULTY OF ARTS.

This Day, 28th of February 2014. Under Supervisor: Dr. Ogugua Paul.

INTRODUCTION

Prior before the advent of the colonial masters, the Igbo people like every other tribe, lived within the confines and comfort of their culture and norms.[i] The indigenous traditionalists believed in the earth goddess, deities and ancestral spirits under a creator called *Chukwu, Obasi, Chi* or *Chineke* (the supreme God). Their beliefs overwhelmed their culture and social lives. One of these beliefs is encapsulated in the word *Ikpenkwumoto*, meaning to judge uprightly. Thus the Supreme Court of Nigeria in **Dabierin v The State** took judicial notice of this cherished custom when she asserted that evidence (testimony) of the elders especially on land matters, are mostly true.

Again the Igbo's cherished brotherhood, communality and relationships. This is buttressed in the word *Umunna, Obinwanne, etc.* Odimegwu recaptures it when he wrote 'The African communalist family engendered dialogue and consensus as the mode of relations and method of governance in the traditional society' (2007:298).[ii] Ogugua summarizes these cardinal virtues of the Igbo community into 'Life, offspring, wealth, truth, *justice, love and peace*' (2003).[iii]

However beautiful and desirous that state of nature was, the dark fog of segregation lingered with it. A certain set of clan or clans are regarded as the 'unclean', 'cult-slave', 'living sacrifice', 'outcast', 'slave of the gods', even 'the untouchables'. Living with them or marrying from them is highly forbidden. Such an act may even convert the defaulter (*Diala*) into an Osu.

Surprisingly, the segregation that befell these unfortunate victims, is not characterized by hatred (for foods are given to the gods knowing the Osu's eat from it, and arms are given to them), rather it culminated from a parochial and fanatical awe of the long aged belief of the 'living sacrifice'.[iv]

EXPLICATION OF TERMS
OSU

This refers to a certain person, clan or species of people in the Igbo community, who are seen as sub-humans or unclean class or as some may say 'slaves to the gods'.

Some communities like *Nzam* in Onitsha, calls it *Adu-ebo*. It is called *Nwani or Ohu-alusi* at Augwu. Some refers to it as *Ume, Ohu, Oru, Ohu Ume...* and some calls it *Omoni (Okpu-Aja)*.

The concept of Osu is not alien to any part of the world. Man's inhumanity to man is a common theme. Thus while we use the Igboland as a case study, we are by no means suggesting that it exist only Igboland. Rather, the appellation – Osu, must bear a generic meaning associated with any form of stigmatization, segregation, victimization, vandalization or Ostracization in any part of the world, without prejudice to any special difference.

THE CULTURE

According to Opata, the concept of culture has undergone a long evolution and has, therefore, acquired an elasticity of meaning such that everything under the sun can be subsumed under the umbrage of culture; hence we have mass culture, elite culture, scientific culture, techni-culture, yam culture, dance culture etc. (Ndianefoo, 2009:137).[v]

Ndianefoo went on to warn that the variety of meaning which culture has acquired should not deter us from probing further for roots original meaning. In fact he traced the origin of the word to Latin, meaning 'cultivation of the soil' but in its metaphorical sense, and according to the Encyclopedia of Philosophy, it means 'cultivation of the mind'. It was during the 18th and 19th where the word extrapolated to include 'beliefs, ideas, attitudes, artifacts, etc.' No wonder Wiredu pontificated that culture is a complete phenomenon including everything that is connected with a people's way of life (Ndianefoo 2009:138).

The cultivation of the Igbo's mind on Osu has existed before the advent of the colonial masters and Christianity. It would therefore not be wrong to say that Osu-ism was and still is the culture of the Igbo man. What we would rather be concerned with, is the conflict (if any) between its culture of love and communality against the Osu tradition; the justification if any; and also the conflict of such tradition with our Law.

THE LAW

According to Okunniga, quoted by Sanni, 'Nobody including the lawyer has offered, nobody including the lawyer is offering, nobody including the lawyer will ever be able to offer a definition of law to end all definitions'(2006:25-26).[vi] This was why Thurman Arnold declared that 'obviously, Law can never be defined'.

Without prejudice to the Austinian definition of law or the Acquinian conception of the Idea law, this study will limit itself to the fundamental human right law that are apt for our present discourse as contained in the Constitution of the Federal Republic of Nigeria.

THE ORIGIN OF OSU CASTE IN IGBOLAND

The true origin of the emergence of the Osu caste seems to be at large. Different stories are told about this living tale. For instance, Amadife tells us that the origin is traced to the era when the gods were believed to demand for human sacrifice during festivals, so as to cleanse the land of abomination. Then the people would contribute to the general purse for a purchase of a slave or for kidnapping one. This victims and their descendants became known as *'Osu arusi'*.[vii]

For Ezekwugo, the origin is traced to the Nri Kingdom (the acclaimed ancestral home of the Igbo man). It is believed that the Nri's possessed a hereditary power and thus do go about cleansing the various kingdoms of abomination. Any community that refutes to be cleansed are dabbed 'osu's' or 'untouchables' (1987:10).[viii]

Some believed they were descendants of travelers who were merely allowed to stay in the community. Others say they are bastards from non-Osu's (*Diala*).

Finally, the stronger view seem to lean on ostracization. This occurs when a particular person or group refutes the decision of the King or the entire community. The people naturally begins to withdraw from the defaulter (this was a traditional method of punishment/criminal justice in the pre-colonial era). Sometimes, the king banishes the defaulter from the land. Upon the passage of time, from one generation to another, the victim or the children of the victim are then referred to as Osu's together with their descendants.[ix]

CONFLICT BETWEEN OSU-ISM AND THE IGBO CONCEPT OF LOVE, UNITY AND COMMUNALITY

The Igbo community is once known for its belief in love, unity and communality. For instance, ownership of land was communal and not individualistic. Liberty was cherished and there was nothing like kingship at its inception. *Anyi nile bu ofu* (we are one) was the emblem. This was why the appointment of warrant chiefs in the indirect-rule system of Lugard led to its failure in the East. Igwe reports that one of the fundamental constitution of the Igbo society and culture is the spirit of liberty.

No one community or village would want to oppress the other. No Igbo man would want to slavishly serve another under normal circumstances. The parable says it all - *Egbe bere ugo bere nke si beya ebele nku kwaa ya* (let both the kite and the eagle perch and stand and let the one that stands in the way of the other become powerless) – (1991:143).[x]

Ogugua further recaps the cherished principle of relationship, belongings, solidarity and the common good.[xi] Aghamelu, in explaining the current clog on solidarity noted:

> The present situation in Africa (Igbo) has created a dualistic society… the weak and powerful. The worst still is that instruments of the state are used to foster this kind of discrimination, a flagrant abuse of the common good. The principle of solidarity is both Christian and African to the core (2003:85).[xii]

However Igwe noted that the spirit of liberty led to individualism and self-deceptive competition (1991). Perhaps, this was what intensified the ambivalent attitude towards eradication of the Osu crisis.

The puzzle here is, how come a community of such values and morals should abhor the opposite of it values? Simply put, can belongings and solidarity co-exist with segregation and ostracization? This may be referred to as 'Identity Crisis' in the words of Oraegbunam (2006:237).[xiii] It is sad indeed to see us making a cultural change by adopting the bad sides of westernism, while we are ambivalent towards accepting the good sides, namely – Abolition of Osu-ism!

CONFLICT BETWEEN OSU CASTE AND THE LAW

Right to freedom from discrimination was provided for by **section 42** of the Constitution. The Constitution frowns at any discrimination of a person on grounds of the person's community, ethnicity, and place of birth or origin, circumstance of birth, sex, religion, political opinion or disability. What is not clear is – **does the law frowns on the other person's right not to want to middle or involve with an *Osu*?** The cases of *Nzekwu v Nzekwu* [1989] 2 NWLR (pt 104) p.373 SC; *Mojekwu v Mojekwu* [1997] 7 NWLR (pt 512) p.282 CA, are to the effect that one's entitlement cannot be denied on basis of discrimination. But surely, this does not extend to forcing a man to associate with another. For the right to associate includes the fundamental right not to associate. And if there is a right not to associate, then to an extent, there exist a right to discriminate. But this discrimination is attenuated and curtailed when a right of the other person to obtain something arises. In other words, in strict legal sense, an Osu as a right to go anywhere any man has right to go; to speak where any man has right to speak; to contest any post any man has a right to contest; and to propose to any lady any man has right to. But these rights are all subject to the rights of those at the other end i.e. the lady, the people to be ruled, the owner of the place etc. while they cannot deter him to apply, he cannot force them to accept. The solution may therefore lie in ethical persuasion.

The right to peaceful assembly and association is provided for in **section 40.** Association includes political parties, trade unions, or any association. However, the section limits this right to the dictates of the Independent National Electoral Commission with respect to political parties. By implication, the section forbids unlawful assembly and associations. Now the Osu's right to associate with the people of his village (which is a lawful community), cannot be abrogated by the fact that the community does not want him. Although one may argue that such right is dependent on the willingness of the other (community) to be associated with. On that regard I am forced to admit that this law is not breached. It will only amount to ethical and moral considerations.

It is notorious that an Osu is prohibited from even coming into the gathering of the *Diala* (freeborn or non-Osu), not to consider him addressing them. His right to be heard is consequently denied. The right to freedom of expression and the press is guaranteed by law. In *Adewole & ors v Jakande*[xiv] it was held that **Section 39** affirms the freedom of expression to every individual and the press. Thus within the provisions of any act enacted by the national assembly, a person is free to own and operate any medium of the dissemination of information, ideas and opinions. Hence the right of press group is ensured too. So is the right of an Osu.

The Right of liberty is affirmed by **section 35 (1) CFRN**, but this Right is made limited to orders of a court i.e. when the accused is found guilty. A person who has not attained the age of 18, may be restricted for the purpose of his education or welfare. Also persons suffering from *infections or contagious diseases,* unsound mind, drug addiction, alcohol or vagrants, may be restricted for the purposes of their care or the protection of the community, or for the purpose of preventing unlawful entry into Nigeria or effecting expulsion from Nigeria.[xv] A lawful arrest upon reasonable suspicion of having committed a criminal offence and contempt of court may be made. Personal liberty has been defined by Prof. Dicey as 'the right not to be subjected to imprisonment, arrest and any other physical coercion, in any manner that does not admit of legal justification'. It must be noted that the *Osu* does not suffer from any physical contagious disease. It is imaginary and socially afflicted. Thus it is illegal within the ambit of the law.

A PHILOSOPHICAL EVALUATION OF THE OSU CASTE JUSTIFICATIONS

Onwubuariri tried to justify the Osu caste by implication when he classified the types of Osu to include:

1. The voluntary

2. The involuntary and

3. The mass consensus classification.

While the last two are not the fault of the victim, Onwubuariri justified the voluntary type of Osu which reflects the victim's choice to become an Osu. This type occurs when the victim, out of laziness, takes to the shrine and eats from the food of the gods. It is also voluntary according to him, when the victim resorts to the shrine for solace out of the frustration or marginalization witnessed as a *Diala*[xvi]. Thus if a man want to be an Osu who are we not to respect that?

With greatest respect, I do not think psychologists nor humanists, will concur with Onwubuariri's implied postulation. Frustration of a maximum depth can indeed cause one to act independent of one's will (which should be a defense under section **24 and 28** of our **criminal code**). Thus it could lead to a natural mental infirmity, which psychologists would label 'Abnormality'.

Another purported justification seems to have its root in Aristotle's conception of equality. Aristotle believed that equals should be treated equally. The implication of this postulate is that we are not all equals. But does the veracity of Aristotle's postulation extend to discuss at hand? No! I do not think so. The initial stage for status scrambling must and should be a fair and equal platform. It is upon the success or failure of one's prowess that the later status should be determined. But then again, the ethical implications are too alarming to ignore.

Moreso, on the argument for punishment; as necessary as punishment may be, ostracization raises the ethical question – *should an innocent B suffer from the crime of A?* I do not think so either. Why should the descendants suffer from their father's deviance? Besides, the difficult dialogue between the corrective or punitive justice systems is awakened! If the people have claimed to be bound by one constitution, then they should refrain from taking the laws into their hands!

Again, some have argued that the Osu discrimination is divine as it is mandated and exemplified by God himself. The book of **Genesis chapter 3,** marked the first banishment and ostracism of Adam and Eve. We all, today, suffer from that wrong. Lucifer himself was banished by God and he suffers the ever-labeled name of *Satan*. He is likened to an *Osu*. What is the difference between and Osu and the then *Gentiles*? The stigmatization of the *Samaritans*? The Christians and the Jews? And yes! Why is segregation common to all parts of the world?

The above arguments, except one, can be dismissed by the simple logical truth – that the world is doing it doesn't make it right. Otherwise, why is the world gradually reversing? Why did the nation agree to shun discrimination? What informed the *Magna Carta?* Why did Nigerians applaud the anti-gay law but frown at the punishment stipulated for defaulters?

As to the biblical argument against God himself, I am forced to humbly delve into the spiritual, for the spiritual cannot be comprehended with the physical alone. Therein lies the age-long border between the Rationalists and the Empiricists. First, let me say there was no ostracization, but a mere banishment of Satan and Man. The book of **Job chapter 1 verse 6**, records the devil coming into the meeting with the sons of God.

That meeting must be periodical and for the devil to attend and have his place and spoke with God (a thing any Osu will die to witness), then Satan is nothing near an Osu in the sense of the word. Neither is man rejected by God, less **John 3:16** would not have existed.

Secondly, every sin can only be punished based on its *quantum meruit*. This is both ethical and natural. If sin is sin, then the punisher will be grossly unfair. In fact some armed-chair Christians have argued that God would have terminated Satan's life than allow the present conflict. Many have died, and the punishment for death ought to be death. This is to say that Satan's punishment is either suitable or alleviated – a perfect ethical justice. As for man, the rule was *if you want to stay here, don't do this.* Man failed. The natural consequence is *you can't stay here.* But because the offspring must not suffer, and the purity of the innocent offspring to stay on board has been contaminated and stolen, Christ had to die to redeem not only the innocent but to give chance to the banished to purify himself and come back. It is on this note that I commend certain rumored but unnamed villages which allows a ritual cleansing of the Osu to come back into the fold. Although the unanswered question would be, will it ever be the same?

CONCLUSION

From this research, it is found that Osu caste still exist within the Igboland, some are well pronounced some are whispered. It is also prevalent in all parts of the world under the disguise of some other names. While the law prohibits segregation of these victims, the law does not and must not mandate forced relations.

The solution to the acceptability of the Osu's, lies more on ethical and moral persuasion. Seminars in conjunction with the local government and the village chiefs and heads should be made to sensitize the people against the dreaded impart of Osu caste. The world is gradually sinking into the Hobbessian state of nature. But this state of nature is false and caused. The

John Locke's state of nature must be redeemed. We must begin afresh to value the true virtues, less we all shall fall. And what a fall would be there my countrymen!

REFERENCES

[i] I. Okodo, 'The Peoples and Cultures of Nigeria' in N Ojiakor (ed), *Salient Issues in Nigerian History, Culture, and Social Political Development.* (Enugu: Emmy-Angel Publishers, 2007).

[ii] I. Odimegwu, 'From Past to Our Present: In Search of Responsible Leadership' in I. Odimegwu et al (eds), *Philosophy, Democracy and Conflicts in Africa* (Awka: Fab Educational Books, 2007) Vol. 2.

[iii] P.I Ogugua, *The Septenary Nature of Igbo Cultural Values* (Double Pee Comm, 2003).

[iv] V.E Dike, *The Osu Caste System in Igboland: Discrimination Based on Descent.* Retrieved from www.nairaland.com/370741/osu-caste-system-igboland-discrimination. On 25/02/2014 at 8am.

[v] IJ Ndianefoo, 'The Role of Philosophy in Nurturing and Sustaining National Culture' in N Okediadi et al (eds), *Themes In Nigerian Peoples and Cultures* (Awka: School of General Studies Unizik, 2009).

[vi] AO Sanni, 'Law in Social Context' in A Sanni (ed), *Introduction to Nigerian Legal Method* (2nd edn, Ile-Ife: OAU Press, 2006).

[vii] Amadife, *'The Culture That Must Die'* Sunday Times, March 23, 1988.

[viii] C.M Ezekwugo, *Ora-Eri Nnokwa and Nri Dynasty* (Enugu: Lengon Printers, 1987).

[ix] M. Okonkwo, from Mmiata-Anam Anambra State. Interviewed on 19/2/2014;
C. Anietu, In Asaba, Delta State. Interviewed on 20/2/2014;
C. Ohum, Awka, Anambra State. Interviewed on 25/2/2014;
L. Njoku, from Owerri, Imo State. Interviewed on 25/2/2014.

[x] Igwe S.S.N, *Social Ethical Issues in Nigeria* (2nd edn, Obosi: Pacific Publishers, 1991).

[xi] Ogugua P, 'Septenary Principles at The Foundation of Igbo-African Communalism' in *Nnamdi Azikiwe Journal of Philosophy,* Awka. (2007) Vol. 1, No. 1.

[xii] Aghamelu F, 'African Development: The Contribution of African philosophy' in C.B Nze

(ed), *Ogirisi: a new journal of African Studies.* (Amawbia: Lumos Nig. Ltd, 2003) Vol. 1, No. 1.

[xiii] Oraegbunam I.K.E, 'Western Colonialism and African Identity Crisis: The Role of African Philosophy' in IKE Odimegwu (ed), *Philosophy and Africa* (Amawbia: UNESCO, 2006).

[xiv] (1981) 1 NCLR 262.

[xv] *Odogu v A.G Federation* [1996] 6 NWLR (pt 456) p.508 SC; *Madiebo v Nwankwo* [2002] 1 NWLR (pt 748) p.426. CA; *Shugaba v Minister of Internal Affairs* [1981] 2 NCLR 459.

[xvi] Onwubuariri F, *Appraising the OSU Caste System in Igboland within the Context of Complementary Reflection,* http://www.frasouzu.com/ Retrieved on 20/02/2014 at 10pm.

ABOUT AUTHOR

Emmanuel Okonkwo is a male Nigerian citizen who resides in the city of Lagos. Born on the 9th of October 1986. He hails from Anambra State, from Mmiata-Anam (Anambra west). He did his secondary school in Amuwo Odofin High School, mile 2 Lagos. He has a bachelor degree in Philosophy and in Law. He is to be called to the Nigerian Bar in November 2014. He taught in various secondary schools. His love for knowledge and writing has shaped his desire to be an author. This led to the publication of his co-work 'Whispers from the Desert', which was reviewed by Orient Newspaper on February 13, 2014. He has written many articles and music as a choral master (under the pen name- Citadel O.I) and won several local awards.